ρ3

The Lost Theory
of
Asclepiades of Bithynia

The Lost Theory
of
Asclepiades of Bithynia

J. T. VALLANCE

CLARENDON PRESS · OXFORD
1990

Oxford University Press, Walton Street, Oxford OX2 6DP
Oxford New York Toronto
Delhi Bombay Calcutta Madras Karachi
Petaling Jaya Singapore Hong Kong Tokyo
Nairobi Dar es Salaam Cape Town
Melbourne Auckland
and associated companies in
Berlin Ibadan

Oxford is a trade mark of Oxford University Press

Published in the United States
by Oxford University Press, New York

British Library Cataloguing in Publication Data
Vallance, J. T.
The lost theory of Asclepiades of Bithynia.
1. Ancient Greek medicine
I. Title
610.938
ISBN 0–19–824248–4

Library of Congress Cataloging in Publication Data
Vallance, J. T.
The lost theory of Asclepiades of Bithynia / J. T. Vallance.
p. cm.
Includes bibliographical references (p. 149)
1. Asclepiades, of Bithynia, ca. 130–ca. 40 B.C. 2. Medicine,
Greek and Roman. 3. Medicine—Philosophy. 4. Philosophy, Ancient.
I. Title.
R126.A72V35 1990 610'.938—dc20 90-33667
ISBN 0–19–824248–4

Set by Joshua Associates Ltd., Oxford
Printed and bound in
Great Britain by Bookcraft (Bath) Ltd.,
Midsomer Norton, Avon

For
Geoffrey Lloyd

Preface

A LITTLE book with so many footnotes probably needs some justification. Mine is that much of the material discussed here is rather obscure and not readily accessible. (Even today a vast amount of later philosophical and medical literature remains effectively unexplored.) Generally I have not sought to provide exhaustive references to ancient works lest the footnotes strangle the book itself. Texts central to my argument are both reproduced and translated. Where a translator is not named, the translation is my own. Minor texts are simply reproduced and relegated to notes.

In what follows, many of my debts will be quite clear. Geoffrey Lloyd first pointed me in the direction of the Hellenistic doctors, and has since helped me at every stage. I owe a very considerable debt also to David Sedley and Vivian Nutton, who read several drafts with great care and tact. Jonathan Barnes has given me much valuable help with my collection of the ancient testimonia. I also owe a great deal to David Johnston, Catherine du Peloux Menagé, and Malcolm Schofield. David Wilson of the *Thesaurus Linguae Graecae* in California, John Scarborough, and the late Elizabeth Rawson helped me get started on Asclepiades. Some of this work was done as a research student at St John's College, Cambridge. For financial support I must thank the University of Sydney, St John's College, Cambridge, and the Cambridge Faculty of Classics, not to mention my parents. A Fellowship at Gonville and Caius College has given me the chance to work in most congenial and stimulating surroundings. I owe special thanks to the officers of Oxford University Press, and in particular to Dr John Waś, who edited a tortuous typescript with great accuracy, saved me from many errors, and gave me the chance to correct more. Needless to say, I am entirely responsible for those which remain.

J. V.

Gonville and Caius College
Cambridge
March 1990

Contents

Abbreviations

CMG   Corpus Medicorum Graecorum (Leipzig and Berlin)

CML   Corpus Medicorum Latinorum (Leipzig and Berlin)

DK   *Die Fragmente der Vorsokratiker*, ed. H. Diels and W. Kranz (6th edn.; Berlin, 1951–2)

Dox.   *Doxographi Graeci*, ed. H. Diels (Berlin and Leipzig, 1879)

DSB   *Dictionary of Scientific Biography*, ed. C. C. Gillispie (New York, 1970–80)

K   *Claudii Galeni Opera Omnia*, ed. C. G. Kühn (Leipzig, 1821–33; repr. Hildesheim, 1964)

LSJ   H. G. Liddell, R. Scott, and H. S. Jones, *A Greek-English Lexicon*, 9th edn. with Supplement (Oxford, 1968)

PTKA   πρὸς τὸ κενούμενον ἀκολουθία

PTLP   πρὸς τὸ λεπτομερὲς φορά

PW   *Paulys Real-Encyclopädie der classischen Altertumswissenschaft*, ed. G. Wissowa *et al.* (Stuttgart, 1894–1980)

SM   *Claudii Galeni Pergameni Scripta Minora*, ed. J. Marquardt, I. Müller, and G. Helmreich, 3 vols. (Leipzig, 1884–93; repr. 1967)

Introduction

This is a philosophical detective story. At the end of the second century BC, a Greek doctor in Rome was making radical changes to the practice of medicine. With a wry twist on Plato's idea of the purpose of philosophy, this man had described traditional Greek medicine as a 'preparation for death'. His own therapeutic practice became famous for its emphasis on humane treatments—non-violent exercise, the regulated use of food, wine, bathing, and massage—and he sought explicitly to distance himself from the more disruptive, dangerous pharmacological and surgical procedures which mark much of ancient medicine.

This doctor was called Asclepiades.[1] His medical ideas were grounded on an extraordinary and mysterious theory. He thought that the body (and indeed the universe) was constructed out of fundamental particles, known in Greek by the equally mysterious name ἄναρμοι ὄγκοι. These particles percolated through the Asclepiadean body. Their balanced, regular movement was characteristic of health; their impaction caused disease.

This small book is an attempt to reconstruct the details and affiliations of the theory out of contradictory accounts of it which were offered in later antiquity. Quite simply, it is an attempt to discover what the ἄναρμοι ὄγκοι might have been. Where did Asclepiades' ideas come from? What was a doctor

---

[1] Biographical and bibliographical details, a general account of his medical practice, and a survey of the ancient evidence relating to him are forthcoming in my chapter on Asclepiades in *Aufstieg und Niedergang der römischen Welt*, II. 37/i. I am also preparing a complete collection (with translation) of the ancient fragments. Rawson (1982) provides the best modern account of Asclepiades' life and date; Rawson (1985) puts him in his historical context. Cocchi (1758) is still important.

doing with a bizarre theory like this? And why did the theory
vanish, apparently without trace? Why did Galen feel the need
to crush it without mercy? The answers to these questions have
important implications for our understanding of Hellenistic
philosophy, as well as medicine. More generally, they relate to
one of the most intriguing doctrinal disputes in ancient science.

Asclepiades came from the town of Prousias-on-the-Sea, in
the (then) Roman kingdom of Bithynia. Next to nothing is
known of his life, beyond the likelihood that he spent a good
part of it working in Rome. He was almost certainly dead by 91
BC.[2] None of his works survive, and he is known to us solely
through the testimony of later, mostly unsympathetic witnesses.
Historically, he stands at a crucial point in the development of
Greek medicine, between the great third-century Hellenistic
doctors Erasistratus of Ceos and Herophilus of Chalcedon,[3] and
Galen of Pergamon, who lived several hundred years later.

He was highly successful—as anyone who advocates the therapeutic benefits of 'passive exercise' is bound to be. Roman
gravitas was outraged in some quarters, and the elder Pliny
seems to be reproducing typically Roman attitudes when he
claims that Asclepiades was merely a jumped-up teacher of
rhetoric who turned to medicine for all the wrong reasons:

The ancient system of medicine remained established, and claimed
considerable traces of its acknowledged province, until the time of
Asclepiades, a teacher of rhetoric in the age of Pompey the Great. He
had not found that career sufficiently lucrative but had a mind that
could look beyond the Forum so he suddenly changed his job. He
started to practise medicine. He had neither experience of medicine,
nor understanding of those remedies which must be learnt through
autopsy and experience; in spite of this he was able to win people over
by his overpowering, daily practised oratory. He abandoned tradition.
Reducing the whole art to the study of causes, he made medicine conjectural. He posited five main types of common therapeutic aids:

---

[2] This date is provided by Cicero. He has Crassus mention Asclepiades in the perfect
tense in the *De oratore* (1. 14. 62), and the dramatic date of this piece is 91 BC. See Rawson
(1982).
[3] A full modern account of Erasistratus is still awaited. For Herophilus, see now von
Staden (1989).

abstention from food, in other cases from wine, body-massage, walking, and rocking. Since everyone realized that these measures were readily available, they were widely approved, as if the easiest things are also the right ones. Asclepiades almost brought the entire human race around to his point of view; it was as though he had been sent from heaven. He used to seduce the minds of his patients with meretricious tricks. Sometimes he prescribed wine (administered at the right time), and sometimes water. Herophilus before him had studied the causes of diseases, and Cleophantus had developed the idea of treatment with wine, so Asclepiades, as Marcus Varro tells us, preferred to be known as the 'cold water giver'. He dreamt up other treats for his patients too—suspended couches, whose rocking he prescribed to alleviate diseases or to induce sleep; baths (capitalizing on men's insatiable greed for such things), and many other congenial, pleasant treatments. He had a great reputation, and the fame to go with it. When once he came across the funeral procession of someone unknown to him, he ordered the man removed from the pyre, and saved him.[4]

Pliny's cynicism is anything but inflexible. Elsewhere he gives a slightly more positive account:

But the greatest fame belongs to Asclepiades of Prousias, who founded a new sect, spurned the envoys and propositions of Mithridates, discovered a way by which wine might be used to treat the ill, brought back a man alive from his funeral and saved him, and who made the mightiest of pacts with fortune—that he should not be thought a doctor if he was ever ill at all. He won his pact; a very old man indeed, he died after falling down stairs.[5]

If cases like this say more about Pliny's method of composition than anything else, Asclepiades' fellow Greeks were not uniformly impressed either. Galen, much later, regarded him as a complete charlatan. Yet Cicero speaks warmly of him and frequently so does Celsus. A little more recently, Elizabeth Rawson described him as one of the leading intellectual figures of the later Republic.

The work of Barnes, Lloyd, Frede, and Smith in particular has underlined the philosophical sophistication of Greek medicine

---

[4] *Naturalis historia* 26. 12–15.
[5] Ibid. 7. 124.

in the age of the medical sects. I am not concerned here so much with tracing the philosophical pedigree of later medical ideas as with examining the more immediate intellectual context in which they developed. Hippocratic doctors had, of course, given a great deal of time to the development and criticism of method and epistemology. That much is clear even from a glance at the first chapters of *On Ancient Medicine*, or *On the Nature of Man*. Hippocratic medicine too is characterized by polemical and rhetorical treatises, especially such sophistic show-pieces as *On the Art* and *On Breaths*. But Hellenistic and later Greek doctors seem to have organized themselves (or in a few cases, been organized) into more or less clearly defined doctrinal groups. Some, like the so-called 'Empiricists', rejected the theoretical study of physiology and pathology altogether; they based their rejection on complex epistemological arguments which owe much to various hues of philosophical scepticism.[6] The Methodists eschewed even the complex epistemological arguments—or so they claimed—and argued that all diseases could be reduced to two or three evident 'phenomenal states' which the doctor could quickly be trained to recognize. *Ars brevis vita longa*, they said mischievously. The Dogmatists were not a cohesive group, as such; later commentators included in their number such unhappy companions as Erasistratus, Asclepiades, and Galen. But the idea behind the category, if we put it simply and perhaps a little naïvely, was that these people accepted the importance of theoretical physiological and pathological research for the doctor. Put not so simply, 'Dogmatist' (or the rather more euphemistic 'Rationalist') was a term of abuse. Galen, by his own account, stands outside these groups, though others do not view his lofty disdain of sectarian doctors who came before him quite so simply as that. Yet much of Galen's writing directed against figures like Erasistratus and Asclepiades is in fact aimed at con-

---

[6] We are almost wholly dependent for our knowledge of medical empiricism on three works by Galen, the *De sectis ad introducendos*, *De experientia medica*, and *Subfiguratio empirica*. These and other ancient testimonia are collected in Deichgräber (1965). Frede (1985) translates the treatises by Galen and with Lloyd (1987: 158–67) gives the best introduction to the epistemological background of the medical sects. The most straightforward ancient account is that of Celsus in the proem to bk. 1 of the *De medicina*.

temporary rivals. Questions of Galen's practice and philosophical affiliations loom large in what follows.

Most of the medical writers I shall be discussing do, unlike Galen, belong to reasonably clearly defined groups. The Methodists include Thessalus of Tralles (*fl.* AD 60), Soranus of Ephesus (*fl.* AD 120), and Caelius Aurelianus (*fl. c.* AD 450). Of Thessalus, little is known; Soranus is well known as the author of the *Gynaecia*, and ultimately of the treatises *De morbis acutis* and *De morbis chronicis*, which survive only in the much later Latin paraphrastic versions of Caelius Aurelianus.[7] Other witnesses for Asclepiades' thought, like Sextus Empiricus (*fl.* AD 190), or Celsus (*fl.* AD 40), may have been doctors, but are better known in other areas—Sextus as a sceptic philosopher and Celsus as an encyclopaedist and savant. In all, nearly forty ancient authors preserve information about our man. Trying to fit together a coherent model of his thought from the intellectual detritus of so many is not easy. The result is a fairly detailed piece of historical and philosophical commentary. I hope that the implications of the book's conclusions will go some way towards justifying its detail. Asclepiades' theory offers a fascinating and unexpected example of how medical theories of matter were developing alongside those of the great Hellenistic schools of philosophy, and how doctors were making original contributions to ancient physical theories to some extent independent of the philosophers.

J. V.

---

[7] All those who use Caelius need at the outset to establish an attitude to his status as a source. My own view is somewhere between that of Smith (1979), who effectively treats Caelius and Soranus as equivalent, and Pigeaud (1982), who perhaps goes too far in crediting him with originality.

I

What were the ἄναρμοι ὄγκοι?

Asclepiades' physiological theory has exercised the minds of scholars sporadically for several hundred years. German minds especially: in Germany he was seen surprisingly often as a precursor of more modern corpuscular theorists.[1] The ancient scholarly and not so scholarly minds, upon which we must rely so heavily, were fiercely divided over the theory's value, and over the niceties of its constitution. Polemical minds (and ancient doctors were well endowed) had their own reasons for discussing theories they disagreed with, and these reasons rarely included the desire to provide us with transparent historical accounts. But the ancient witnesses for Asclepiades (and by this I shall mean, somewhat arbitrarily, those who wrote before AD 1000) are agreed that he explained disease in terms of the interaction of 'theoretical' and constantly moving elemental corpuscles in theoretical passages in the body. The corpuscles are variously called ὄγκοι, ἄναρμοι ὄγκοι, στοιχεῖα, *corpuscula*, and *moles*; the pores πόροι, ἀραιώματα, *foramina*, and *viae*. Impaction, also given a number of names by the different ancient witnesses, including ἔμφραξις, ἔνστασις, *statio*, *obtrusio*, or *coacervatio*, was the immediate cause of many diseases. Certain other diseases, it seems, were not explained in this way, and they formed a different category of affections. For want of a more neutral term, I shall refer to this theory as Asclepiades' corpuscular hypothesis.

That much is uncontroversial. In the reconstruction of the hypothesis which follows, I shall begin with an examination of

[1] e.g. by Burdach (1800) and Lasswitz (1879).

what these strange corpuscles might have been. The second part of my argument, in Chapter 2, covers the role of the pores in Asclepiades' system, and how these informed a principle called 'movement towards what is fine' or πρὸς τὸ λεπτομερὲς φορά, which Asclepiades deployed to explain the movement of fluids in the body. The final part of the study is an attempt to understand Asclepiades' motivations, affiliations, and subsequent influence.

The philosophical history of Greek corpuscular theories of matter is long and distinguished. The systems which come most readily to our minds are those of Leucippus, Democritus, and Epicurus. These philosophers posited a universe made up of unalterable, totally massy 'atoms', and their negation, void. Yet from the fifth century onwards a plethora of alternative corpuscular hypotheses were on offer. The traditional (and slightly unhelpful) distinction which Aristotle makes, for example at the beginning of the *De generatione et corruptione*, between those philosophers we now tend to call 'Monists' and the 'Pluralists' brackets the early atomists rather uneasily together with the likes of Empedocles and Anaxagoras.[2] These last two seem to have envisaged a universe composed out of some kind of plurality, which in the case of Empedocles underpinned his four-element theory, and for Anaxagoras involved the homoiomeries. Plato too had what amounts to a corpuscular model of matter, which reduced the universe to primary geometric shapes. A considerable amount of attention was focused on issues such as the possibility of division with finite limits (a necessary basis for Democritean atomism), but these corpuscular theories can also be distinguished from each other by the solutions they offered to the problem of giving elements room in which to move. The atomists' solutions to the problems of motion through the supposition of void are well known. For Democritus, void was a three-dimensional, spatial extension, the negation of 'being', yet with a somewhat nervous existence in its own right. Epicurus emboldened the concept, by effectively licensing the right of

---

[2] e.g. at *De generatione et corruptione* 314^{a}1–314^{b}1.

non-existence to existence. But those who denied the possibility of three-dimensionally extended void within the universe (the so-called 'continuum theorists') were far from silent. For some, like Plato in the Timaeus, void could not exist except as the material substrate of body. The space which Plato's elemental geometric shapes occupy was necessarily 'incorporeal' in every sense of the word, and without any physical extension beyond the limits of the bodies which occupy it.[3] Plato's models lived on in the form of enhancements and developments at the hands of such figures as Heraclides of Pontus and Xenocrates of Chalcedon.[4] Over in the Lyceum, Strato of Lampsacus, Theophrastus' successor, was a particularly interesting case—he did not adhere to an atomistic conception of void, but none the less seems to have held that void extensions in three dimensions can exist in nature in the form of interstices between ill-fitting particles of matter. He went further too, and almost certainly argued that a full-scale three-dimensional void space could be created artificially, by force, using suction equipment.[5] By the beginning of the third century BC, these rival models had been joined by the hard-line continuum theory of Stoic physics, which held that void space simply does not exist anywhere within the world.

In Asclepiades' day, these problems were still contentious. Many doctors were acquainted with the details of such disputes amongst natural scientists and philosophers. Erasistratus' physiological theory, we will see, shows clear signs of influence from Strato, and perhaps from the Platonic tradition. Asclepiades too needs to be understood against the background of the work done by philosophers. But he was first and foremost a doctor, and it is from this point that I shall begin my investigation.

Most modern attempts to come to grips with the theoretical

---

[3] *Timaeus* 53 C–61 C.

[4] Heraclides was a pupil of Plato, and a member of the Academy from *c.*360 to 339 BC; Xenocrates of Chalcedon was head of the Academy from 339 until 314 BC.

[5] Modern work on Strato begins effectively with Diels (1893). Like Asclepiades, Strato is known only through the testimony of others, and there is considerable disagreement over the details of his physics. The main ancient texts relating to Strato are collected in Gottschalk (1965). I shall have more to say about Strato in ch. 2.

details of Asclepiades' pathology have made the description of
the corpuscles their first objective. My attempt is also modern,
though it should be noted that this kind of approach in the past
has been informed by the assumption that Asclepiades is more
interesting for the light he casts on his predecessors than he is
for his own thought. The work of Wellmann, Harig, and Rawson
this century has done much to show that Asclepiades was an
original physician and philosopher in his own right, but he is still
regarded in some quarters as a rather cranky source for
Heraclides of Pontus, with whom he is sometimes connected in
the ancient doxographies.[6]

Scholars have produced a bewildering array of interpretations
of the confused and contradictory evidence. Put very broadly,
there are three influential lines of thought about the ἄναρμοι
ὄγκοι and their intellectual ancestors. The first and most
dogged sees Asclepiades as a medical atomist, and the corpuscular hypothesis as an adaptation of Epicurean atomism.[7] Its most
vocal modern supporter is Pigeaud, who argues for the strong
thesis that the corpuscular hypothesis is so patently an atomistic
one that Asclepiades cannot properly be understood outside the
Epicurean tradition.[8] There are others disinclined to go so far,
but who are keen not to exclude such a link altogether.[9] Epicurean atomism, after all, became the supreme corpuscular
system in later antiquity. And Pigeaud has a powerful ally in
Galen, who draws no substantive distinction between the Epicurean and Asclepiadean theories. Galen cannot be brushed
aside today any more easily than he could in antiquity. Some

[6] Wellmann (1908), Harig (1983), and Rawson (1982). One of the most important
recent accounts, that of Gottschalk (1980), does not allow Asclepiades any real intellectual independence from Heraclides at all.

[7] This is the standard position found in the older histories of science and medicine,
from which I select Alpinus (1611), 8; Haller (1776), 141; Blumenbach (1786), 49; Zeller
(1909), iii. 569–72; Reymond (1927), 90; Castiglioni (1947), 199–200; Berghoff (1947),
24–5; Singer and Underwood (1962), 51; Scarborough (1969), 38.

[8] Pigeaud (1981a), 141. Pigeaud also suspects an anti-Asclepiadean backlash in
Lucretius.

[9] e.g. Allbutt (1921), 189 n. 2, Neuburger (1910), i. 202–3. Furley and Wilkie (1984),
38–9, say that Asclepiades was an atomist, and leave it at that. Harig (1983) tries to keep
his options open, accepting a role for Heraclides and Epicurus in the development of
Asclepiades' thought.

doctors in later antiquity even saw Epicurus in Asclepiadean terms. This is especially true of the later commentators on Galen, who accepted the Epicurus–Asclepiades connection without ever assessing it critically. Most spectacular perhaps is the comment of the anonymous Paris scholiast on Galen's *De elementis secundum Hippocratem*, who implies that Epicurus called his atoms ἄναρμα: οἱ περὶ Δημόκριτον λέγοντες τὰς ἀτόμους, Ἀναξαγόρας δὲ ὁμοιομερείας, ὁ δὲ Ἐπίκουρος ἄναρμα, 'the Democriteans call them atoms, Anaxagoras calls them homoiomeries and Epicurus *anarma*'.[10]

Gumpert, the first editor of the ancient testimonia relating to Asclepiades, devoted a large amount of space to an exposition of Epicurean atomism as he found it in Lucretius, before suggesting that the Asclepiadean model might well have been assimilated to the Epicurean (admittedly, he said, to an unknown extent) as a result of the 'carelessness of the scribes'. Asclepiades' corpuscles, said Gumpert, were fragile, and this was an insuperable obstacle to their being the same as Epicurean atoms.[11] Fragile? But how? This suspicion was to become an institutionalized worry (rightly so) amongst some of those who were prepared to concede that in other respects Asclepiades took Epicurean ideas on board.[12] It led to the growth of a second body of opinion, notably supported by Sprengel in his *Versuch einer pragmatischen Geschichte der Arzneykunde* of 1792. Sprengel argued that the doxographical reports linking Asclepiades' corpuscular hypothesis with Heraclides of Pontus had not sufficiently been taken into account. He maintained that whatever the links with Epicurus, Heraclides was very likely to have been behind at least some aspects of the theory.[13] Even though Sprengel did not argue that Asclepiades followed Heraclides slavishly, it has been widely imagined that the systems of Asclepiades and Heraclides

---

[10] Parisinus Gr. Suppl. 634, fo. 20. At Galen, *De theriaca ad Pisonem* 14. 250 K, Epicurus is assimilated into the company of doctors. Heidel (1909) suspected this kind of 'reverse contamination' without finding any clear examples of it.

[11] Gumpert (1794), 59, sect. 41. This problem had also been picked up by Le Clerc (1702), ii. 107, and later by Tourtelle (1804), 403–4.

[12] e.g. Tourtelle (1804), 404.

[13] Sprengel (1792), ii, sects. 6–7, esp. p. 12.

were similar, if not identical. And this is why much of the recent research on Asclepiades has been done by scholars who are mainly interested in Heraclides.[14]

An examination of Voss's collection of the ancient Heraclidean testimonia will reveal that there are *no* non-doxographical references to Heraclides' corpuscular theory which do not also include Asclepiades. Since the two (I shall argue) had entirely different reasons for adopting such a theory, it is naïve to suppose that the theory's medical manifestations in Asclepiades need cast any accurate light on how the non-medical Heraclides had used it—if he had used it at all in the form we know. In fact, Harig has recently put forward some compelling arguments for suspecting that there were significant doctrinal differences between Heraclides and Asclepiades.[15] Lonie and Gottschalk too provide balanced accounts of this position—they are careful not to say too much about Heraclides on the basis of evidence explicitly relating only to Asclepiades, and their work remains fundamental to any account of the subject.[16]

The problems of explaining a theory by appealing to another, earlier theory which is even less well understood are obvious. Yet this criticism could be directed at another influential position (or, more strictly, group of positions) which grew out of Diels's work on the physics of Strato of Lampsacus.[17] Diels had argued for a link between Strato, Erasistratus, and Asclepiades. His argument was grounded on the belief that each of these men adopted variants of the principle of horror vacui. This, in Diels's view, indicated a conception of the nature of void rather different to that found in the atomists.

The great Wellmann took exception to all this. He also ques-

---

[14] Lasswitz (1890), 212–14, and Heidel (1909) argued that the theories of the two were practically identical; cf. Gottschalk (1980), 38–9, and Bäumker (1890), 325, who traces the theory from Empedocles to Heraclides to Asclepiades. He is followed by Horne (1963), 325, and Guthrie (1965), 150 n. 1. Stannard in *DSB* s.v. Asclepiades wisely hedges his bets.

[15] Especially on the question of Heraclides' medical interests (or lack of them): Harig (1983), 48 ff.

[16] Lonie (1964; 1965) and Gottschalk (1980), 19–20.

[17] Diels (1893).

tioned the orthodoxy that Asclepiades was a founder of medical atomism. Wellmann initially argued for a tradition which began with the *Timaeus* and then progressed via the shadowy forms of Philistion and Chrysippus the doctor to Erasistratus, and finally to Asclepiades via a certain Aegimius of Elis.[18] He was later to shift his position towards implicating Heraclides.[19] More recently, the picture has been complicated further. Diels's work on Strato has come in for a certain amount of criticism; in particular, the idea that the preface to Hero of Alexandria's *Pneumatica*—hitherto regarded as a crucial source—is transparently Stratonic has been challenged in some quarters. Schmekel has argued that Hero was an atomist, and therefore closer to Asclepiades than to Strato. Gatzemeier has argued that one of our other major sources for Strato, Simplicius, did not understand Strato's physics at all. Few modern scholars would agree with Gatzemeier, but Furley has recently attacked the tendency to attribute to Strato what cannot be laid at the door of Democritus and Epicurus.[20]

In an influential paper which appeared in 1929, Wellmann argued that there was considerable Democritean influence in the Hippocratic corpus. Given the extraordinary influence of the Hippocratic writings, this had obvious implications for students of the philosophical background to later medical theories. Many thought Wellmann had gone too far, and for a long time his views about Democritus and the Hippocratics were out of favour,[21] but the idea of a long-term Democritean influence on medicine has been resurrected most recently by Stückelberger. Committing himself more than most have done, he traces a line directly from Democritus through Erasistratus to Asclepiades and on to Lucretius.[22]

[18] Aegimius has still not yielded his secrets, and probably never will. Philistion of Locri was a 4th-cent. doctor who was influenced by the medicine of the *Timaeus.* Chrysippus (not the Stoic of the same name) was a teacher of Erasistratus.
[19] The direct implication of Heraclides comes at Wellmann (1908), 695.
[20] Furley (1989). Culprits include Schmekel (1938), 110, and Gatzemeier (1970), 94-7.
[21] They were attacked most recently (and sensibly, in my view) by Lonie (1981*a*), 113 ff., (1981*b*), 128.
[22] Stückelberger (1984), esp. pp. 101-16 on Asclepiades' place in the tradition. For the background to Stückelberger's remarkable position, which seems to me to make little of

So much for the Moderns. What of the Ancients? Sextus
Empiricus is the earliest (around AD 190) surviving witness to tell
us that Asclepiades called his primary corpuscles ἄναρμοι ὄγκοι.
Even then, Sextus uses the phrase only thrice;[23] elsewhere he
calls them στοιχεῖα θραυστὰ καὶ ποιά, 'fragile elements endowed with qualities',[24] or 'theoretical ὄγκοι'.[25] The hypothesis has, however, come to be referred to as the 'theory of the
ἄναρμοι ὄγκοι'. Sources earlier than Sextus speak simply of
some kind of corpuscular theory which formed the basis of a
pathological system. This they never fully explain, generally
because of their uniform hostility to it. Dioscorides, for instance,
attacks a group of followers of Asclepiades in the preface to his
Materia medica because he sees the corpuscular theory as an
ultimate threat to received systems of pharmacological explanation:

It needs to be said that although the amount of information handed
down by the ancients was small, at least they were accurate. We must
not agree with recent authorities like Julius Bassus, Niceratus,
Petronius, Niger, and Diodotus—followers of Asclepiades to a man.
They thought it quite reasonable to include information which is
common and known to everyone. (And even then only accurate up to a
point.) They dealt with the properties and testing of drugs only briefly.
The activities of drugs were not tested experimentally. Instead they
carried on inanely about causation, and described each drug in terms
of differences in the corpuscles.[26]

Soranus, the great gynaecologist, attacks the theory because it
leads to the idea that there are no diseases specific to women—
women, after all, are made out of the same elemental corpuscles
as men.[27] Soranus' hostility, we shall see, had a still more human
origin; he had implicitly taken on board a number of Asclepi-

___

the distinction between different types of ancient atomism, see his (1972), (1974), and
(1979).

[23] *Adversus mathematicos* 9. 363, 10. 318, and *Pyrrhoniae hypotyposes* 3. 32.
[24] *Pyrrhoniae hypotyposes* 3. 33.
[25] *Adversus mathematicos* 3. 5 ἐν λόγῳ θεωρητοὶ ὄγκοι; 8. 220 νοητοὶ ὄγκοι.
[26] Dioscorides, *Materia medica*, pref. 2.
[27] Soranus, *Gynaecia* 3. 3.

adean concepts himself, and had good reasons for keeping quiet about them. Celsus, ever the moderate Roman gentleman, mentions Asclepiades in a list designed only to point out that differences in the therapeutic systems of doctors can be related to their divergent theoretical backgrounds. According to Celsus, Herophilus had argued that all diseases have their cause in 'the humours', Erasistratus in the transference of blood into the arteries which brings about first inflammation, then fever. Asclepiades, he continues, explained disorders by appealing to a blockage of corpuscles which in the healthy body percolate through invisible pores in the body. He goes no further than that:

neque esse dubium quin alia curatione opus sit si ex quattuor principiis vel superans aliquod vel deficiens adversam valetudinem creat, ut quidam ex sapientiae professoribus dixerunt; alia, si in umidis omne vitium est, ut Herophilo visum est; alia, si in spiritu, ut Hippocrati; alia, si sanguis in eas venas quae spiritui accommodatae sunt transfunditur et inflammationem, quam Graeci φλεγμονή⟨ν⟩ nominant, excitat, eaque inflammatio talem motum efficit qualis in febre est, ut Erasistrato placuit; alia si manantia corpuscula per invisibilia foramina subsistendo iter claudunt, ut Asclepiades contendit. eum vero recte curaturum quem prima origo causae non fefellerit.

[Those doctors who advocate the importance of studying the hidden causes of disease say that] there is no doubt that one course of treatment will be needed if ill health is the result of some kind of overabundance or deficiency of the four elements, as some philosophers believe, and another if every morbid condition has its origin in the humours, as Herophilus believed. Yet another type of treatment is called for if illness is caused by breaths (Hippocrates' view), and another if blood transfused into the vessels designed to carry pneuma gives rise to inflammation, which the Greeks call *phlegmone*— inflammation which brings about the kind of movement associated with fever, as Erasistratus thinks. And yet another again if corpuscles which percolate through invisible pores block their passage by piling up, as Asclepiades maintains. The man who has got the prime origin of the cause correct will discover the correct method of treatment.[28]

[28] Celsus, *De medicina* 1, proem 14–15.

Scholars today tend to see the explanation of what ἄναρμος means as both the key to, and most problematic part of, the inquiry into the exact nature of the theory. The term is certainly a difficult and rare one. When its use is specifically attributed to named individuals, it is *only* in connection with Heraclides and Asclepiades. The word also occurs twice in Philostratus' third-century-AD work on gymnastics.[29] It can be found at least seventeen times in the Greek Galenic corpus, in a variety of configurations and often in cases where Asclepiades is not mentioned in the immediate context.[30] Nowhere does Galen make any connection—implicit or explicit—concerning the use of the word between Heraclides of Pontus and Asclepiades. The term is always applied by Galen to elemental particles, and there seems no doubt that Asclepiades is always in his mind even when he is not actually there in the text. (That is not the same as saying that Galen *only* has Asclepiades in mind.)

My own discussion of what the word might mean will occupy much of what immediately follows. It is important though, to have some idea of what the ὄγκοι might be. ὄγκος is a common word in Greek, and it carries a huge variety of meanings. To take just a few cases which might at first seem relevant: Aristotle uses it (in a non-theoretically loaded way) to mean 'bodily bulk' and in one place opposes it to 'void'.[31] It reappears with a roughly similar sense in a Democritean testimonium, from Aristotle's *De caelo*.[32] ὄγκος also appears in later doxographies. Lonie has

---

[29] *De gymnasticis* 29, 48. The word is used to describe weak hip-joints in men unfit for athletic exercise.

[30] Galen uses the word in the following forms: τὸ ἄναρμον (*De sanitate tuenda* 6. 15 K, *De constitutione artis medicae* 1. 249 K, cf. *De elementis* 1. 417 K); ἄναρμα . . . στοιχεῖα (*De elementis* 1. 416, 500 K, *De naturalibus facultatibus* 2. 39, 98, 102 K, and cf. ἄναρμα καὶ ἀμέριστα στοιχεῖα at 2. 100 K); σώματα ἄναρμα (*De naturalibus facultatibus* 2. 101 K; cf. *De usu partium* 4. 350 K in Helmreich's Teubner edn., where Kühn reads ἀμεριστῶν for ἀνάρμων; ἄναρμα (*De dignoscendis pulsibus* 8. 923, 927 K, and cf. *De morborum differentiis* 6. 840 K, *De methodo medendi* 10. 852 K, *Commentarium in Hippocratis de natura hominis librum* 15. 36 K, *Commentarium III in Hippocratis librum III Epidemiarum* 17A. 506 K). Vivian Nutton tells me that the Laurentian MS of the fragmentary *De substantia facultatum naturalium*, at 4. 761 K, also preserves the term in the list of particles there. The word occurs only once in what we might call the ancient 'ps.-Galenic' doxographical corpus, at *De historia philosophica* 18.

[31] *Physics* Γ 203ᵇ28-31.

[32] *De caelo* 305ᵇ1-16 = 68 A 46a DK.

warned that cases where the use of the term applied to elemental particles is attributed by late authorities to early philosophers are suspect because they may have been contaminated
by Asclepiades' own usage.[33] We should be aware of this, even if
we do not wish to go so far as to suggest that Asclepiades had
this kind of influence over the doxographers.[34] Simplicius, who
does not mention Asclepiades in his extensive discussion of
atomic theories, claims to be quoting from Aristotle's treatise *On
Democritus* when he suggests that Democritus called visible
agglomerations of atoms ὄγκοι: νομίζει δὲ εἶναι οὕτω μικρὰς
τὰς οὐσίας, ὥστε ἐκφυγεῖν τὰς ἡμετέρας αἰσθήσεις. ὑπάρχειν δὲ
αὐταῖς παντοίας μορφὰς καὶ σχήματα παντοῖα καὶ κατὰ
μέγεθος διαφοράς. ἐκ τούτων οὖν ἤδη καθάπερ ἐκ στοιχείων
γεννᾷ καὶ συγκρίνει τοὺς ὀφθαλμοφανεῖς καὶ τοὺς αἰσθητοὺς
ὄγκους, 'He [Democritus] considers that the particles are so
small as to escape our senses. They possess all manner of forms
and shapes, and they differ in size. He generates and puts
together the masses which present themselves to our sight and
which are perceptible out of these particles which are effectively
elements'.[35] Similarly, we should be rather wary of Aetius' report
that Empedocles and Xenocrates divided their own elements
into ὄγκοι which were themselves the 'elements of elements',
that is to say of earth, air, fire and water: Ἐμπεδοκλῆς καὶ
Χενοκράτης ἐκ μικροτέρων ὄγκων τὰ στοιχεῖα συγκρίνει, ἅπερ
ἐστὶν ἐλάχιστα καὶ οἱονεὶ στοιχεῖα στοιχείων, 'Empedocles and
Xenocrates compose the elements from still smaller masses—
from *minima* in fact, which are rather like the elements of
elements'.[36] Yet the word is very common in Epicurus, as a brief
glance at Arrighetti's index will show. Epicurus does not

[33] Lonie (1965), 131 n. 4.
[34] Sextus Empiricus, *Adversus mathematicos* 10. 43, could conceivably lend weight to Lonie's suggestion; here we have a case of someone apparently familiar with Asclepiades' pathology calling the Epicurean atoms (or molecules) ὄγκοι. But this is not enough on its own.
[35] Simplicius, *De caelo*, pp. 294-5 Heib.=68 A 37 DK; cf. Diogenes Laertius 9. 44. Even if the usage is Democritean (and that, of course, is open to doubt), the word need not have a technical sense here.
[36] Aetius, *Placita* 1. 17. 3=31 A 43 DK.

consistently apply it to his atoms specifically, but uses it rather more vaguely to mean a 'discrete quantity' or 'bulk'.[37]

I could cite many other examples, but in a sense they are beside the point. The word has a separate medical pedigree, and it is this which really informs the background to Asclepiades' usage. These senses can be illustrated most appropriately and conveniently in Galen. In one sense, related to the verb ὀγκόω (='swell up'), the word is used for 'lumps' in the body generally, and sometimes tumours in particular.[38] Elsewhere, an ὄγκος is the swelling of the arm above a tourniquet,[39] and 'size' generally.[40] In Galen's account of Erasistratus' theory of respiration, it is used for the 'volume' of air we breathe.[41] At the beginning of his treatise *De tumoribus praeter naturam*, it seems that Galen is only prepared to apply the term to visible magnitudes. Soon we shall see that his position is not altogether consistent.

Only a few ancient witnesses shed any direct light on the nature of Asclepiades' corpuscles, and I am necessarily bound to them, initially at least. Caelius Aurelianus is the one authority to summarize the theory for its own sake (even if his own purposes lurk not too far away in the background). He does this in the first book of his Latin version of Soranus' treatise *De morbis acutis*; I reproduce the text below, together with an unusually detailed apparatus, to convey some idea of the directions taken by the modern textual work. Imaginative textual criticism has flourished in the absence of any useful surviving manuscript tradition prior to the first printed editions. The text can, and has been, read in a variety of different ways, and no amount of textual criticism seems capable of removing some basic ambiguities and indeterminacies.

---

[37] So Lonie (1964), 160 n. 12. Pigeaud (1981*a*), 174 (with n. 134), disagrees with me, and uses the Epicurean examples as a corner-stone of his argument that Asclepiades was an Epicurean. With the more general use of the term compare Sextus Empiricus, *Adversus mathematicos* 7. 287 ὁ γὰρ ἄνθρωπος οὐδέν ἐστι παρὰ τὸν ὄγκον καὶ τὰς αἰσθήσεις καὶ τὴν διάνοιαν.

[38] e.g. ὄγκοι παρὰ φύσιν; *Ars medica* 1. 335 K, *De atra bile* 5. 116 K, *De causis symptomatum* 7. 106 K. The Clementine *Recognitions* in Rufinus' translation (8. 15) glosses *oncos* with 'tumores' or 'elationes', in the context of Asclepiades' theory.

[39] *De anatomicis administrationibus* 2. 387 K (ὀγκώθη). Cf. 2. 433 K.

[40] *De usu partium* 3. 50 K of the tendons, 3. 880 K of the tongue.

[41] *De usu respirationis* 4. 473 K.

Asclepiadi responsuri eius primum dogma proponamus, qua voluti apprehensionis falsitate peccatis etiam involvuntur curationum. primordia namque corporis primo constituerat atomos, [secunda] corpuscula intellectu sensa sine ulla qualitate solita, atque ex initio comitata, aeternum moventia. quae suo incursu offensa mutuis ictibus   5 in infinita partium fragmenta solvantur magnitudine atque schemate differentia; quae rursum eundo sibi adiecta vel coniuncta omnia faciant sensibilia, vim in semet mutationis habentia, aut per magnitudinem sui, aut per multitudinem, aut per schema, aut per ordinem. nec, inquit, ratione carere videatur quod nullius faciant qualitatis corpora: aliud   10 enim partes, aliud universitatem sequetur. argentum denique album est, sed eius affricatio nigra; caprinum cornu nigrum, sed eius alba serrago.

1 voluti] veluti *Rm*   3 secunda] *om. edd. post G: an* secundo *Drabkin in not.*
5 comitata] commutabilia *Wellmann apud Susemihl (1892), 432 n. 81*: incompacta
Wellmann (1908), 695 n. 2: concitata *Voss (1896), 66*: compacta *Pigeaud (1981b), 197*
aeternum moventia] a. se m. *Wellmann (1908), 695*   10 videatur] videtur *R*

Caelius' statement of Asclepiades' doctrine begins at section 105, and continues beyond the end of this quotation, through to section 115. Although he is hostile to the theory, Caelius' refutation does not begin until rather later. The Latin is difficult, and the progression of Caelius' thought not easy to determine. Before I offer my own translation, I should explain how I arrived at it (and perhaps apologize for the nature of the explanation). The first major problem surrounds line 3. Should *secunda* stand? If so, what does it mean? Drabkin's suggested emendation *secundo* rests on the assumption, mistaken I think, that the word is connected syntactically with *primo*. While this would be consistent with one aspect of Caelius' usage, *primo* could just as easily have an absolute sense, as it generally does when Caelius is introducing a new section.[42] It would be more natural here, perhaps, to translate *primo* as 'originally' or 'initially'. *Secunda* then would not have to be a corresponding connective. The sentence begins emphatically with *primordia*; *secunda* occupies a similar position at the beginning of the next clause. If there is to

[42] As at *De morbis acutis* 1. 10 'primo ad aliud ex alia re transire videbimur'; 1. 11 'primo in his quae decimo scripsimus enodavimus'; 1. 116 'contradicens igitur primo incusat clysteres'—just to take a few examples from the immediate context.

be any correspondence, this is where we would expect to find it. It might be reasonable to see in this some kind of contrast between *primordia* and *secunda corpuscula*. This is in fact how Gottschalk, one of the most recent commentators, reads the sentence. 'If it [*secunda*] is kept,' he argues, 'it would imply that the *corpuscula* are a second order of particles less fundamental than the *atomi*; the latter would presumably be identical with the *fragmenta* described in the following sentence.'[43] Gottschalk's interpretation has much to recommend it. But I do not agree that it is based on the most natural reading of Caelius. The lack of a connective between *atomos* and [*secunda*] *corpuscula* could just as easily lead us to expect that the second clause contains an explanation of the first. My translation would run like this:

We are going to answer Asclepiades,[44] but first let us set out the doctrine which leads him and his followers[45] into their therapeutic blunders. Firstly, Asclepiades posited atoms as first principles, [second] corpuscles perceptible to the intellect and without any normal quality, gathered together from the beginning and in constant motion. When they strike each other with mutual blows as a result of their particular kind of conflux, they are dissolved into innumerable fragments of parts differing in size and shape. They come together again, and through their addition and conjunction create all sensible things. They have in them the power of change in respect of their size, number, shape, and arrangement. It is not illogical, says Asclepiades, that bodies with no quality should make up the sensible world. For one thing is true of the part, and another of the whole. So it is that silver is white, but a sliver of silver is black; goat's horn is black, but a shaving of it is white.

Other problematic aspects of this translation can wait. What is significant at this stage is that it would appear that the *corpuscula* can themselves be divided further. How far? An answer to this question first means finding out what ἄναρμος means.

Purely etymological attempts to pin down the meaning of the word have been unsatisfactory, although Heidel came remark-

---

[43] Gottschalk (1980), 57.

[44] Caelius has been discussing Asclepiades' treatment of phrenitis.

[45] The change from singular (*eius*) to plural (*involvuntur*) is common in Caelius in this type of context; it suggests that Caelius has the Asclepiadeans as well as Asclepiades in mind here.

ably close, in my view, to diagnosing the correct sense. Examining the meaning of related terms has been of little value.[46] The possibility that Heraclides of Pontus used the word does not cast much light on how a doctor might have employed it.[47] There is little philological doubt as to what it *should* mean, but that is beside the point.[48] How are we to understand its application to elemental particles?

According to Caelius, the corpuscles can be dissolved into fragments as a result of their mutual collisions. The present tense *solvantur* points to a continuing tendency innate in the corpuscles, just as do the reflexive pronouns. The particles *are prone* to being broken up: they must therefore correspond to the θραυστὰ στοιχεῖα of the Greek reports.[49] The verb Caelius uses to describe this breaking up is *solvo*; the *corpuscula quae solvantur* could then be termed *soluta*. I do not believe that it has been noticed before that *solutus* (and *solubilis* in later Latin) is used with senses very close to those attested for ἄναρμος.

Heidel[50] diagnosed that in the two cases in Philostratus where the word ἄναρμος is applied to an athlete's hip-joint,[51] the sense required is 'weak' or 'loose'. With this one might compare the parallel use of *solutus* in Latin: for example, Petronius, *Satyricon* 140. 6 'podagricum se esse lumborumque solutorum ... dixerat', and Pliny, *Naturalis historia* 31. 59 'est autem utilis sulpurata [sc. aqua] nervis, aluminata paralyticis aut simili modo solutis ...'. In both these cases, the word evidently means 'weak' and it is a

---

[46] Gottschalk's account at (1980), 38 ff., is probably the most useful. Cf. Heidel (1909), which is still important. As Gottschalk points out, the word is made up of the privative ἀν- and ἁρμός.

[47] Heraclides may well have written a dialogue called περὶ τῆς ἀπνοῦ— *On the Woman who Stopped Breathing*—and, yes, he is mentioned by Galen along with Asclepiades in one passage in the *De tremore, rigore et palpitatione* (7. 615-16 K). But the dialogue does not prove that Heraclides was particularly interested in medicine. And Galen is simply noting that the Pneumatist physician Athenaeus 'had something to say' about Heraclides, Asclepiades, and Strato of Lampsacus. *Pace* Lonie *et al.*, we have no real idea of the context of Athenaeus' mention of them.

[48] It should mean something like 'jointless'.

[49] As at ps.-Galen, *Introductio sive medicus* 14. 698 K; Sextus Empiricus, *Pyrrhoniae hypotyposes* 3. 32-3.

[50] Heidel (1909), 19-20.

[51] *De gymnasticis* 29, 48.

general, qualitative description of a constitution rather than of a specific visible defect. There is the idea here that something is lacking which should normally hold the parts of the body firmly together. Another usage, which at first looks slightly different, but turns out to be quite useful for my argument, comes in the Younger Pliny (*Epistulae* 4. 2. 3): 'habebat puer mannulos multos et iunctos et solutos'. The term here is the opposite of 'yoked', 'harnessed'. Still more prosaic than ponies, perhaps, is Cicero, *Partitiones Oratoriae* 53: '[verba] ... soluta, quae dicuntur sine coniunctione, ut plura videantur'.[52]

If the term *solutus* were to be applied to an elemental particle, we might imagine that the particle in question would be liable to shatter into fragments because it lacks internal cohesion. *Solutus* and *solubilis*, then, are very likely to be Latin equivalents of ἄναρμος; certainly they share the attested and predicted meanings of the Greek word. By way of a preliminary and provisional conclusion, one might suggest that ἄναρμος must mean something like 'weak' or 'loosely held together'. This is almost exactly what Lonie had in mind. Heidel (1909) thought that the word was synonymous with θραυστός: ἄναρμος could certainly carry a hint of this sense too.

This interpretation also solves a textual problem in Calcidius' *Commentary on Plato's* Timaeus. At ch. 215 of the commentary, Calcidius is speaking of the earlier philosophical views of Asclepiades, Democritus, and Epicurus on the material constituents of the soul:

aut enim moles quaedam sunt leves et globosae eaedemque admodum delicatae, ex quibus anima subsistit, quod totum spiritus est, ut Asclepiades putat, aut ignitae atomi iuxta Democritum, qui ex isdem corporibus et ignem et animam censet excudi, vel id ipsum atomi casu

---

[52] Cf. the tradition of quasi-atomistic language and analogies in Greek grammatical theory. Dionysius of Thrace, for example, gives us a nice case in Greek when he refers to the διάλυσις of words into their component letters at *De compositione verborum* 14. For some more examples, one might note Prudentius, *Hamartigenia* 504–5 'nam vanum quidquid sol aspicit, ex elementis | cuncta solubilibus fluxoque creamine constant.' Also Prudentius, *Peristephanon* 10. 501–10, where *solubilis* is applied to the body in a quasi-medical context. Ammianus Marcellinus 16. 8. 10 mentions a portable, modular bridge which he calls a *pons solubilis*.

quodam et sine ratione concurrentes in unum et animam creantes, ut Epicuro placet, ob similitudinem atomorum, quarum una commota omnem spiritum, id est animam, moveri simul.

Is the soul (which is entirely *pneuma*) made up from light, round lumps, which are very frail, as Asclepiades thought, or out of the atoms of fire, as Democritus believed? Democritus thought that fire and soul were fashioned from the same bodies. Or do atoms, running together by chance and without reason, create the soul, as Epicurus thought? Because of the similarity of the atoms, when one is moved, the whole of the *spiritus*, that is the soul, is moved at the same time.[53]

Clearly enough, the 'lumps' (*moles*) are to be identified with the corpuscles. Even so, it is an unusual word to choose; it is generally used of larger masses, and it occurs in none of the other Latin witnesses.[54] I understand *globosae* as 'round', but with some trepidation. Surely Asclepiades' particles cannot all have been spherical; this would make a nonsense of all the other reports which give them a variety of shapes. Some fancy philological footwork might yield the sense 'clustered together', thus giving much the same sense as *comitata* in the Caelius passage.[55] But then again, Calcidius is talking about the particles which make up the soul, and it is their very uniformity that gives the soul its character.

Most important, however, is *admodum delicatae*. I follow Waszink here, although *deligatae* also has good manuscript authority. Indeed, in the context of bonds and connections *deligatae* ('tied together') might seem at first sight the right word. But to read *admodum deligatae* would effectively be to stress the integrity and the strength of the fine corpuscles which make up the soul. And this flies in the face of all the other witnesses to the corpuscles' fragility. *Delicatus*, on the other hand, shares several

---

[53] Calcidius, *In Timaeum* 215, pp. 299–30 Waszink.

[54] Clement of Rome, for instance, translates ὄγκοι as 'tumores vel elationes', clearly following the medical sense of the word, and this is an interpretation favoured by most of the later Latin translators of Galen.

[55] I cannot offer any firm ancient parallels for *globosus* with this sense, but see *OLD* s.vv. *globosus* b, *globo* 2, where there are some examples of the verb applied to bees swarming (from the elder Pliny). Note also the use (in Tacitus) of the word for a 'throng of people'. I have to admit that the only other occurrence of *globosus* and *moles* together is at Augustine, *De genesi ad litteram* 1. 12, where *globosus moles* = 'the sun'.

senses with *solutus* in Latin, and, indeed, with the senses of ἄναρ-μος found in Philostratus. A list of examples would be otiose, but here are a couple: the younger Pliny says of Silius Italicus, who died in extreme old age even though he was not ill, that he was *delicato magis corpore quam infirmo*.[56] *Delicatus* could conceivably have a sense close to θραυστός as well. Classical Latinists might like its moral overtones.

This makes sense of a host of troublesome texts. I shall just take two for the moment. In the treatise *De constitutione artis medicae*, Galen declares that οὐ μὴν οὐδὲ τὸ ἄναρμον τὸ Ἀσκληπιάδου θραυστὸν ὂν ὀδυνήσεται θραυόμενον, ἀναίσθητον γὰρ ἔστιν, 'the ἄναρμος particle of Asclepiades, being fragile, will not cause pain when it is broken, since it does not admit of feeling'.[57] The throw-away suggestion 'when it is broken' makes it quite clear: the particles are further divisible.[58] So too in the pseudo-Galenic *Introductio sive medicus*: κατὰ δὲ Ἀσκληπιάδην στοιχεῖα ἀνθρώπου ὄγκοι θραυστοὶ καὶ πόροι . . ., 'according to Asclepiades the elements of man are fragile corpuscles and pores'.[59]

There is, of course, rather more to it than this. If we accept that the *corpuscula quae solvantur* are in fact the ἄναρμοι ὄγκοι, then there is still some way to go. Why, for example, does Caelius use the term *atomus* here, with all its Abderite connotations of indivisibility? This need not be too serious. Just what Soranus' Greek was here is anyone's guess, but he always calls the Asclepiadean particles ὄγκοι, never ἄτομοι. Caelius may be using *atomus* as a Latin word (it is his usual policy to flag all words he has transliterated from Greek), and by this stage in

[56] *Epistulae*, 3. 7. 9. For some other cases see Apuleius, *Metamorphoses* 5. 10 (*delicatus manus*); Lactantius, *De opificio Dei* 12. 12 (*delicati artus*); Caelius Aurelianus, *De morbis acutis* 2. 35 (*delicatae membranae cerebri*).

[57] 1. 249 K.

[58] The suggestion need *not* be that Asclepiades explained pain in terms of the breakage of corpuscles (*pace* Lonie (1965), 127, 139). On the contrary, according to Caelius, *De morbis acutis* 1. 119, pain for Asclepiades is the result of the impaction of large corpuscles in pores. Galen has only brought pain in here because of his conviction that if a body is to be capable of feeling, it must submit to qualitative change, something Asclepiades cannot account for. Lonie cites *De methodo medendi* 10. 852 K, but that passage too is making the same point as Galen is here.

[59] 14. 698 K.

Latin the term could have lost some of its original, specific sense. I have to admit, however, that I have found no cases of *atomus* being applied specifically to non-Abderite or Epicurean particles. My case *may* get some support from Galen's treatise *De experientia medica*, where the corpuscles are described (in Walzer's translation at 24. 6) as 'atoms . . . parts which cannot be divided further'. I suspect that this sentence may contain a gloss by Hunain; Galen nowhere calls the ὄγκοι ἄτομοι, nor does he ever say explicitly that the ὄγκοι are *not* divisible further. Stückelberger takes this passage in Galen as evidence for a kind of Democritean atomism in Asclepiades, but he privileges this rather dubious case over all the others where Galen says exactly the opposite.[60]

There is another related difficulty which needs much closer consideration—were the corpuscles endowed with qualities? Sextus explicitly says that they were—he says they are θραυστὰ καὶ ποιά[61]—while Caelius insists that they are *sine ulla qualitate solita*: 'nec, inquit [sc. Asclepiades], ratione carere videatur quod nullius faciant qualitatis corpora [sc. omnia sensibilia]' ('Nor, says Asclepiades, is it illogical to suppose that all sensible things are created out of bodies with no quality').[62] Galen is often taken to be siding with Caelius. What are we to make of all this?

The philosophical debate over 'quality' has its roots in early Democritean arguments in favour of finite divisibility on the one hand, and the qualitative difference between atoms and the visible agglomerations and qualities made out of them on the other. If it can be shown that Asclepiades is working in the same tradition as this, then *my* fledgeling argument will be in very serious difficulty. This difficulty, among other things, led Gottschalk ultimately to suggest that Asclepiades' theory involved two distinct levels of particles, one with qualities and one without.[63] His argument rests partially on the belief that Caelius'

---

[60] Stückelberger (1984), 110–11.
[61] *Pyrrhoniae hypotyposes* 3. 33. *Adversus mathematicos* 10, 318, often cited here, is slightly different, and I shall consider it shortly.
[62] *De morbis acutis* 1. 105; cf. 1. 106.
[63] Gottschalk (1980), 47 ff.

account is not altogether consistent with the other crucial text relating to Asclepiades' pathology, Sextus Empiricus, *Adversus mathematicos* 3. 5. In that offending passage, Sextus is giving an illustration of what the word ὑπόθεσις can mean:

οὐ μὴν ἀλλὰ καὶ κατὰ τρίτην ἐπιβολὴν ὑπόθεσιν καλοῦμεν ἀρχὴν ἀποδείξεων, αἴτησιν οὖσαν πράγματος εἰς κατασκευήν τινος. οὕτω γοῦν τρισὶν ὑποθέσεσι κεχρῆσθαί φαμεν τὸν Ἀσκληπιάδην εἰς κατασκευὴν τῆς τὸν πυρετὸν ἐμποιούσης ἐνστάσεως, μιᾷ μὲν ὅτι νοητοί τινές εἰσιν ἐν ἡμῖν πόροι, μεγέθει διαφέροντες ἀλλήλων, δευτέρᾳ δὲ ὅτι πάντοθεν ὑγροῦ μέρη καὶ πνεύματος ἐκ λόγῳ θεωρητῶν ὄγκων συνηράνισται δι' αἰῶνος ἀνηρεμήτων, τρίτῃ δὲ ὅτι ἀδιάλειπτοί τινες εἰς τὸ ἐκτὸς ἐξ ἡμῶν ἀποφοραὶ γίνονται, ποτὲ μὲν πλείους ποτὲ δὲ ἐλάττους πρὸς τὴν ἐνεστηκυῖαν περίστασιν.

Furthermore, in a third application of the term, we call the starting-point for a demonstration a 'hypothesis', in the sense of a 'postulate for establishing something'. So we say that Asclepiades used three hypotheses in his demonstration of the blockage which causes fever. First, that there are intelligible pores in us, differing in size from each other, second that parts of moisture and pneuma are gathered together from all sides out of intelligible corpuscles which are in permanent motion, and third, that there are continuous emanations from us to the outside world, which vary according to the prevailing condition.

For Gottschalk there are two distinct types of particle here. The lower corpuscles (the less fundamental) are endowed with qualities, and are represented by the μέρη in Sextus' account, while the higher and more fundamental, correspond to the *fragmenta* of Caelius and have none.[64] It is an ingenious answer to the problem, but not the only one. The Italian scholar Bignone had another idea. Consider the Caelius text once again. The corpuscles break up into innumerable fragments, and these in turn reform to create all sensible things. The fragments (or the

---

[64] Gottschalk (1980), 48–52. Gumpert (1794), 57–62, followed a line quite close to Gottschalk's when he drew an analogy between the *corpuscula* and *fragmenta* of Caelius' report, and the συγκρίσεις and ἄτομοι of Epicurus. Such a comparison is, in my view, difficult to sustain. The distinction between atoms and atomic agglomerations is quite clear in Epicurean physics (e.g. *Ad Herodotum* 40–2; Lucretius, *De rerum natura* 1. 483–4). It is very unlikely that our sources for Asclepiades should be so vague about the corpuscles as to neglect this kind of distinction.

primordia, depending on how we understand the second *quae*) have in them the capacity for change (*vis mutationis*) 'either through their size, their shape, or their arrangement' ('aut per magnitudinem, aut per schema aut per ordinem'). Bignone cunningly saw in *mutatio* a way out of the problem: the capacity for change corresponds to the qualities of the corpuscles in Sextus' report.[65]

The first of Asclepiades' hypotheses is clear enough in the Sextus text. There are theoretical pores of different sizes in our bodies. The second is rather more difficult to pin down. The text suggests to me that Sextus is simply trying to say that bodily fluids and pneuma are made from the corpuscles. The 'parts' (μέρη) are, in my view, to be understood as the corpuscles themselves. There is no real problem with the first part of this supposition. Pneuma, according to Caelius Aurelianus' summary of the Asclepiadean doctrine on digestion, is made up of *corpuscula*:

et neque ullam digestionem in nobis esse, sed solutionem ciborum in ventre fieri crudam et per singulas particulas corporis ire, ut per omnes tenuis vias penetrare videatur, quod appellavit λεπτομερές, sed nos intelligimus spiritum.

Furthermore, there is no such thing as digestion in us, but a raw *solutio* of food forms in the belly, and it passes through the individual parts of the body, apparently penetrating all the fine pores. He calls it λεπτο-μερές but we know it as *spiritus*.[66]

In another place, Caelius goes further:

item aliqui Asclepiadis sectatores gestationes et lavacra et vaporationes cataplasmatum atque malagmatum excluserunt in iecorosis, suspicantes tenuissimorum corpusculorum fore consensum, hoc est spiritus, quem λεπτομέρειαν eorum princeps appellavit, atque in egestorum constrictione falsitate causarum adiutoria magna recusantes.

Again, some followers of Asclepiades excluded the use of rocking, baths, and steaming with cataplasms and plasters in cases of liver

---

[65] Bignone (1940), 190. It is only fair to point out that Bignone is not particularly interested in Asclepiades here.

[66] *De morbis acutis* 1. 113.

diseases, suspecting that a sympathetic affection of the finest corpuscles might arise—that is to say of the *spiritus*, which their leader called *leptomereia*. Held back by the poverty of their causal doctrines, they reject the important treatments.[67]

There is nothing here to suggest that these corpuscles which make up τὸ λεπτομερές are secondary in any way. More on this shortly. Now the third hypothesis: there are continuous emanations from us, whose level is determined 'according to the prevailing condition'.[68] The reference here is most probably to Asclepiades' flux theory, which I will introduce in detail in the next chapter. It seems to involve the idea that natural emanations from the body vary when the paths of the corpuscles are interfered with.

Let us return to the second hypothesis. We would reasonably expect here an account of how the blockage occurs. Gottschalk is not necessarily correct in assuming that the statement is a general one telling us that 'parts of moisture and air [are] aggregated out of ὄγκοι'.[69] On the contrary, Sextus may be making the more specific point that 'parts of fluid and pneuma'[70] are augmented (συνερανίζω—'contribute jointly', LSJ) from all parts of the body (παντόθεν) by theoretical corpuscles which are in constant motion. On this reading, we need not imagine that the ὄγκοι and μέρη are at all different. I would prefer this to the weaker thesis that the μέρη are merely composed themselves out of ὄγκοι; a thesis that receives no support outside this passage.

---

[67] *De morbis chronicis* 3. 65.

[68] The text is opaque to me: τὴν ἐνεστηκυῖαν περίστασιν is translated 'the condition prevailing at the moment' in Bury's Loeb, and by Gottschalk (1980), 45, as 'according to circumstances'. It is not inconceivable that there is a reference here to the ἔνστασις or blockage itself. Compare Galen's use of περίστασις in a summary of aetiologies of inflammation, one of which is certainly attributable to Asclepiades, at *De totius morbi temporibus* 7. 444 K: εἴτε γὰρ ἐπὶ παρεμπτώσει συμβαίνει τὸ τοιοῦτον, εἴτ᾿ ἐπὶ τῇ σφηνώσει πάντων τῶν ἀγγείων, εἴτ᾿ ἐπὶ περιστάσει τινὶ τῶν ὄγκων ἐν τοῖς πόροις. If περίστασις simply were to mean 'condition', it would be far too general a term in this context. Could it be equivalent to ἔνστασις? It is tempting indeed.

[69] Gottschalk (1980), 47.

[70] By this ungainly expression I mean the blood and other fluids which naturally pass through the pores of the healthy body; cf. the *sucorum ductus* at *De morbis acutis* 1. 106.

What looks then like a set of very general statements can in fact be read as a detailed account of how Asclepiadean 'impaction' comes about. (Sextus was a doctor, after all.) This is, as it turns out, a fairly typical example of an Asclepiadean disease aetiology. Sextus has no particular interest in outlining Asclepiades' theory beyond its usefulness in illustrating the unacceptable nature (for a Pyrrhonian sceptic) of scientific hypotheses. Caelius, on the other hand, does at least have general description as his aim. This is not to deny that there are relations between the passages. On a verbal level, both Sextus and Caelius use a very unusual verb to describe the gregarious nature of the corpuscles. '*Comitata*' in the Caelius text has been widely suspected by textual critics, some of whom have searched for a translation of ἄναρμος lurking somewhere behind it. Wellmann[71] was the leader of such critics. The proliferation of solutions to this non-existent problem seems to have been inspired by the tendency of some Latin translations of Galen (notably those printed by Kühn) to render ἄναρμος into Latin with 'incompacta'. Variations on this theme take the place of *comitata* in many cases. Of *comitata*, Le Clerc said in desperation: 'je n'entens pas ce . . . mot, si ce n'est qu'il ait voulu dire, que les atomes étoient joints les uns aux autres.'[72] A little more recently, Pigeaud[73] discussed some of the evidence before making an extraordinary suggestion: *compacta*. Pigeaud is, at least, prepared to keep *comitata*, but only on condition that it is rendered with an Epicurean sense parallel to συγκρινόμενα. But there is no reason for all this suspicion. συνερανίζω is a colourful term with a basic, active sense of 'contribute socially', and *comitor* is a reasonable Latin translation. It is a distinctive term to apply to the behaviour of elemental particles, and does not seem to occur in this kind of context before Sextus.[74]

---

[71] Especially Wellmann (1908).
[72] Le Clerc (1702), ii. 106.
[73] Pigeaud (1981*b*), 197.
[74] The other case in Sextus, at *Adversus mathematicos* 7. 294–5, may represent a similar usage, although there it is used of a visible agglomeration. There are related examples of this idea applied to particles in Caelius Aurelianus, e.g. *De morbis acutis* 3. 220 'concursus atque congressus corpusculorum'. A number of passages in Caelius also point to the use

The use of *comitor* may have important light to shed on the
way in which Asclepiades' particles combine. Sextus tells us that
the corpuscles are in constant, restless motion, and that when
there is fever, the parts which make up fluids and pneuma are
augmented by more such corpuscles. What is the nature of this
augmentation or 'contribution'? I shall suggest in the next
chapter that part of the answer is contained in the Asclepiadean
theory of 'movement towards what is fine'. Sextus' report does
imply that there is some kind of dynamic principle at work
behind the movement of free corpuscles towards the pneuma
and fluid in the pores. A preliminary answer, however, is
provided by Caelius' parallel use of *comitata*. As far as he is concerned, the *corpuscula* congregate before breaking up on one
another. There is no hint that they actually join physically; they
simply have a natural tendency to collide with one another (this
is the force of *suo incursu*), and when they strike, they shatter.[75]

This is all very well, but it still does not explain why Caelius
said the corpuscles were 'sine ulla qualitate solita' and Sextus the
opposite.[76] The idea that compound bodies can dissolve into
indivisible parts differing only in arrangement looks like quite
reasonable Democritean or Epicurean atomism. Whatever
Sextus may be up to, Caelius is clearly not talking about compounds here. He is talking about *primordia*, about fundamental
particles. We would never expect to hear the Epicurean 'molecu-
lar' agglomerations referred to in such terms.

Caelius' report leaves us with corpuscles devoid of normal
qualities, which are broken down into fragments, similarly without quality. According to Gottschalk, Sextus is saying that the

of *comitor* in more specifically medical contexts: *De morbis acutis* 1. 103 'sed passionem [sc.
phreniticam] magis atque eius *comitantia* consideremus', where the *comitantia* are what
'keep company' with the disease, contributing to its severity. Cf. also *De morbis acutis* 3.
134, *De morbis chronicis* 4. 18, not to mention the use in medical English of 'concomitant'.

[75] Harig (1983), 51, believes that Asclepiades' corpuscles may have had hooks, like
Democritean atoms, but his argument is based on the supposition that the theory of
magnetism outlined by Galen at *De naturalibus facultatibus* 2. 49–51 K (which has atoms
from the magnet and the object attracted interlocking) is Asclepiadean, where it is surely
to be associated with Epicurus. I discuss this problem in ch. 2.

[76] As Lonie (1964), 158 n. 10, quite rightly observed.

parts which make up pneuma have some kind of existence independent of the ultimate corpuscles, while themselves still
being 'parts'. Are we to assume then that dissolution cannot take
place directly from sensible materials to the level of fragments
without passing through the intermediate stage of corpuscles?
That both fragments and corpuscles come in many sizes seems
clear. How then do they differ? My solution to the problem is a
radical one; it springs both from my suggestion that Caelius did
indeed translate the term ἄναρμοι ὄγκοι in his account with the
phrase *corpuscula quae solvantur, and* that the modern debate
about the qualities of the particles is on the wrong track. Let us
consider the texts more closely. First Sextus at *Pyrrhoniae
hypotyposes* 3. 33:

οὐ γὰρ δήπου δυνησόμθα καὶ τοῖς περὶ Ἀσκληπιάδην συγκατατίθε-
σθαι, θραυστὰ εἶναι τὰ στοιχεῖα λέγουσι καὶ ποιά, καὶ τοῖς περὶ
Δημόκριτον, ἄτομα ταῦτα εἶναι φάσκουσι καὶ ἄποια, καὶ τοῖς περὶ
Ἀναξαγόραν, πᾶσαν αἰσθητὴν ποιότητα περὶ ταῖς ὁμοιομερείαις ἀπο-
λείπουσιν.

We cannot agree with the followers of Asclepiades, who maintain that
the elements are fragile and have qualities, *and* with the Democriteans,
who say they are indivisible and without quality, *and* with the followers
of Anaxagoras, who leave every perceptible quality to the homoiomeries.

Then, at *Adversus mathematicos* 10. 318, he notes:

ἐξ ἀπείρων δ' ἐδόξασαν τὴν τῶν πραγμάτων γένεσιν οἱ περὶ Ἀναξαγόραν τὸν Κλαζομένιον καὶ Δημόκριτον καὶ Ἐπίκουρον καὶ ἄλλοι
παμπληθεῖς, ἀλλ' ὁ μὲν Ἀναξαγόρας ἐξ ὁμοίων τοῖς γεννωμένοις, οἱ δὲ
περὶ τὸν Δημόκριτον καὶ Ἐπίκουρον ἐξ ἀνομοίων τε καὶ ἀπαθῶν,
τουτέστι τῶν ἀτόμων, οἱ δὲ περὶ τὸν Ποντικὸν Ἡρακλείδην καὶ
Ἀσκληπιάδην ἐξ ἀνομοίων μὲν παθητῶν δέ, καθάπερ τῶν ἀνάρμων
ὄγκων.

The followers of Anaxagoras of Clazomenae, Democritus, and
Epicurus, and a great host of others, hold that the generation of things
comes about from innumerables. But while Anaxagoras thinks that
generation comes about from things similar to the things generated,
the Democriteans and the Epicureans hold that things are generated

from bodies which are dissimilar and impassive—that is to say, from atoms. Meanwhile, the followers of Heraclides of Pontus and Asclepiades think that generation takes place from bodies like the ἄναρμοι ὄγκοι which are dissimilar to the products of generation, but passive.

So much for Sextus. Is there an alternative to the supposition that there are two distinct types of particle being referred to here? Yes. The disagreement is not about whether or not the particles had 'qualities'. These different witnesses are quite simply not on the same wavelength. The first passage above is the only one where Sextus says that the corpuscles were ποιά. This passage effectively places Asclepiades in a tradition initially characterized by Democritean atomism and the language used to describe it. But, as ever, Sextus is trying to demonstrate how 'dogmatists', who are all ostensibly claiming to have the correct answers, in fact disagree on even the most basic issues. He has a vested interest in making them look as silly as possible. Nowhere else does he repeat this explicit statement that the particles are ποιά. Elsewhere in Sextus they are παθητά—capable of suffering action—but no more. In the light of *Adversus mathematicos* 10. 318, it seems reasonable to understand ἀπαθῶν as 'indivisible' and παθητῶν as the opposite. Then Sextus would be drawing a simple distinction between Leucippus and Democritus on the one hand, and Asclepiades on the other. In effect this ends up not being so much a discussion of the 'qualities' of the particles, as one centred on their ultimate, physical integrity. It seems clear (to me at least) that as far as Sextus is concerned, Asclepiades' particles are παθητά simply because they are breakable, when the Democritean and Epicurean atoms are most emphatically not so. It is tempting to imagine also that this is all he means by saying that they are ποιά.

What then of the cases where Gottschalk believes Galen to be contradicting Sextus over the quality of the corpuscles? I think that Galen's objections to Asclepiades' corpuscular hypothesis have more of a medical than a philosophical pedigree. When he is attacking Asclepiades, Galen tends to be unhappy about two things in particular: the explanation of the sensation of pain, and the nature of attraction within the body. His arguments about the nature of pain, and its sensation, are lifted not from any philosopher, but straight from the Hippocratic treatise *On the Nature of Man*, where there are similar attacks on purveyors of 'Monistic' theories. The philosophical aspect is there, to be sure; the Yale scholiast on Galen's *De elementis secundum Hippocratem* notes (somewhat lamely) that Galen wrote this work because he did not want the doctrine of Epicurus, Democritus, Asclepiades, or any other like them to be promulgated, because their doctrine 'removed' the elements.[77] But this is the Hippocratic Galen speaking, and Galen's philosophy has its first roots in the medical tradition. When it comes to post-Hippocratic corpuscular theories, Galen can be conveniently old-fashioned at times.

In the *De elementis*, for instance, Galen sees the elements in the body as qualities which act on one another not through 'separation and combination' but through actual alteration of other qualities, or through being altered themselves.[78] Any other explanation, argues Galen (after Hippocrates), will simply 'not account for the phenomena'. 'If man were made out of particles, and therefore from one type of substance, he could feel no pain.'[79] The general assault on these kinds of 'unitary' theories is summed up in the essay *De constitutione artis medicae* at 1. 248 K: εἰ ἀπαθές ἐστι τὸ τῆς σαρκὸς στοιχεῖον, οὐκ ὀδυνηθήσεται, 'If the element which makes up flesh does not admit of feeling, then there will be no pain'. The point is subsequently made in more detail:

τοὺς γοῦν δακτύλους εἰ συμπλέξῃς ἀλλήλοις, εἶτ᾽ αὖθις ἀποχωρίζοις, οὔθ᾽ ἡ σύνοδος, οὔθ᾽ ὁ διαχωρισμὸς ὀδύνην ἐργάσεται. τὸ μὲν γὰρ ὀδυνᾶσθαι σὺν τῷ πάσχειν ἐστίν. πάσχει δὲ οὐδὲν τὸ ψαῦον, ἐπειδήπερ ἐν δυοῖν τούτοιν ἐστὶ τὸ πάσχειν, ἀλλοιώσει τε τῇ δι᾽ ὅλων καὶ λύσει

---

[77] *Scholia in De elementis* 3. 24-7, 3. 34-49 Moraux. The elements for Galen are, of course, those of the 'Hippocratic' four-element theory. (See Smith (1979), 86 ff.)

[78] *De elementis* 1. 486 K; for the explicit attribution of this view to Hippocrates see 1. 456 K.

[79] *De elementis* 1. 417-18, 486 ff. K, to take just two examples from the present work.

τῆς συνεχείας. ὁπότ᾿ οὖν ἐν τοῖς παθητικοῖς ἐναργῶς σώμασιν οὔθ᾿ ἡ
σύνοδος, οὔθ᾿ ἡ ἄφοδος, ὀδύνην ἐργάζεται σχολῇ γε ἂν ἐν τοῖς
ἀπαθέσιν ἐργάσαιτο. οὐ μὴν οὐδὲ τὸ ἄναρμον τὸ Ἀσκληπιάδου
θραυστὸν ὂν ὀδυνήσεται θραυόμενον, ἀναίσθητον γὰρ ἐστίν. ὥστε οὐδὲ
τούτῳ πλέον ὀδύνης ἔσται ἐξ ὧν πάσχει τῆς αἰσθήσεως ἀπούσης,
ὥσπερ ὀστῷ καὶ χόνδρῳ καὶ πιμελῇ καὶ συνδέσμῳ καὶ θριξί. καὶ γὰρ
ταῦτα πάντα πάσχει μέν, οὐκ ὀδυνᾶται δέ, διότι μηδὲ αἰσθάνεται.

If you twist your fingers around each other, and then separate them,
their conjunction and separation will not cause any pain. If pain is to be
felt, something must suffer something. Touching does not involve
suffering; suffering consists of these two things, total alteration and the
dissolution of continuous existence. If conjunction and separation do
not cause pain in bodies which admit of feeling, this will hardly be the
case with bodies which do not admit of feeling. So, the fragile particle
of Asclepiades will not generate pain when it is broken, since it is not
sensible. No more will there be pain in the case of bodies which can
suffer something, when feeling is not present—bodies like bone, fat,
cartilage, sinew, and hair. For all these bodies can suffer action without
hurting since they are not even capable of feeling.[80]

Galen's discussion of pain, then, in both the *De elementis* and
the *De constitutione artis medicae*, hinges on the assumption that if
something is to give pain, it must be both sensible and capable of
feeling. One *might* say he is drawing a distinction between
things which are παθητά (capable of being acted upon, like
Asclepiades' fragile particles) and those which are παθητικά
(capable of feeling). Although Galen does not make such a distinction, we could say that Asclepiades' particles are παθητά
because they can be broken, but are ἀπαθής because they are
qualitatively inert. Sextus makes a similar observation at *Adversus mathematicos* 9. 80, using the same kind of example. Bodies,
says Sextus, which are made up of separate parts do not
sympathize (συμπάσχει) with each other. He illustrates this
with the following case: all the soldiers in an army but one can
be killed in battle without this physically harming the sole
survivor. But if this last soldier, made up of parts which do

---

[80] I. 249–50 K.

sympathize, then proceeds to cut his finger, this will affect his entire body.[81]

Galen has built up a characterization of παθητικὰ σώματα as bodies which admit of feeling. That is as far as he goes. Asclepiades' particles cannot be used to explain sensation. This is because they are like bone, cartilage, fat, ligament, and hair; one can do something to all of these things, but they are themselves devoid of the capacity to feel it. So far, Galen seems to be showing a blanket disregard for any differences between the 'Monistic' theories he is attacking. He is certainly not worried about the niceties of philosophical talk about quality. We can, however, press him a little further. At *De constitutione artis medicae* I. 248 K he separates the theories 'which make man one' into two groups:

εἰ ἀπαθές ἐστι τὸ τῆς σαρκὸς στοιχεῖον οὐκ ὀδυνηθήσεται· ἀλλὰ μὴν ὀδυνᾶται· οὐκ ἄρ᾽ ἐστιν ἀπαθές. εἰ δὲ καὶ πλείω λέγοι τις εἶναι τὰ στοιχεῖα, μὴ μέντοι γε ἀλλοιούμενα, καὶ ἐπ᾽ ἐκείνων ὁ αὐτὸς λόγος ἐρωτηθήσεται κατὰ τὸν αὐτὸν τρόπον. εἰ ἀπαθῆ τῆς σαρκός ἐστι τὰ στοιχεῖα, οὐκ ἀλγήσει· ἀλλὰ μὴν ἀλγεῖ· οὐκ ἄρ᾽ ἐστιν ἀπαθῆ τὰ τῆς σαρκὸς στοιχεῖα. ὁ μὲν οὖν πρότερος λόγος ἀνατρέπει τήν τε τῶν ἀτόμων, καὶ τὴν τῶν ἀνάρμων, καὶ τὴν τῶν ἐλαχίστων ὑπόθεσιν. κατὰ δὲ τὸν δεύτερον ἥ τε τῶν ὁμοιομερειῶν ἀναιρεῖται δόξα, καὶ ἡ Ἐμπεδοκλέους. καὶ γὰρ οὗτος ἐκ τῶν τεσσάρων στοιχείων βούλεται συνίστασθαι τὰ σώματα μὴ μεταβαλλόντων εἰς ἄλληλα.

If the [single] element which makes up flesh is impassive, then there will be no sensation of pain. But pain is felt. Therefore, the element in question is not impassive. And even if someone says that there are several kinds of element, which do not change, then they too will be begging the same question in the same way. If the *elements* which make up flesh are impassive, then there will be no pain. But there is pain. Therefore the elements which make up flesh are not impassive. The first argument [i.e. the one dealing with one element] overturns the hypothesis of atoms, ἄναρμα, and *minima*; the second demolishes the doctrine of homoiomeries and that of Empedocles. For Empedocles would have it that bodies are made up from the four elements, which do not change into each other.

---

[81] Compare Galen's discussion of sympathy at *De naturalibus facultatibus* 2. 39 and *De usu partium* 4. 350 K.

If Galen is prepared to argue that Anaxagoras' homoiomeries and Asclepiades' ὄγκοι are ἀπαθεῖς then this is surely proof that he is not talking about the same thing as Sextus. The term for Galen is not connected with indivisibility at all—he is simply not interested in *that*—but with being 'feelingless' in the context of physiological explanation. Similarly, at *De morborum differentiis* 6. 840 K, he refers to Epicurus and Asclepiades as proponents of the idea that the body is ultimately made up of one kind of thing. Here the sense which he customarily gives to ἀπαθῆ is quite clear: if the body is made up of atoms or ἄναρμα or some altogether feelingless things, he argues, then qualities will reside in the arrangement of the parts, much like a house built out of 'impassive stones' which is not complete—does not have the quality of 'being a house'—in every arrangement of its parts.[82] This, argues Galen, just as the author of *On the Nature of Man* did before him, is no kind of model to use in explaining physiological phenomena. A cliché, even an outdated argument against atomism, including the Democritean variety, perhaps. But Galen's attention must be understood here against the background of his own ideas about combination, change, and quality; he is telling us nothing about what the Asclepiadean particles themselves were really like. He continues 'if the stones [that make up a house] were able to suffer things' then qualitative variety would be present in the house as a result of more than just varying arrangements.[83] Galen then applies this to medicine, saying that if the elements of our bodies were able to suffer action, then quantitative variety would be exhibited in complete alteration of the parts, rather than mere alteration of arrangement. The very fact that Galen can use the word ἀπαθής to describe stones shows just how far he is from Sextus.

[82] οὐκ οὖν ἕν ἐστι τὸ τῶν ζῴων σῶμα, καθάπερ ἢ ἄτομος ἡ Ἐπικούρειος, ἢ τῶν ἀνάρμων τι τῶν Ἀσκληπιάδου· σύνθετον ἄρα πάντως. ἀλλ' εἰ μὲν ἐξ ἀτόμων, ἢ ἀνάρμων, ἢ ὅλως ἐξ ἀπαθῶν τινων σύγκειται, τὸ μᾶλλόν τε καὶ ἧττον ἐν τῷ ποιῷ τῆς συνθέσεως ἕξει δίκην οἰκίας ἐξ ἀπαθῶν μὲν λίθων συγκειμένης, οὐ μὴν ἐν τῇ συνθέσει γε πάντῃ κατορθουμένης. . . . οὐ γὰρ ἐνδέχεται παθεῖν τι τὴν ἄτομον αὐτήν, ἀλλ' ἐν τῇ συνθέσει τε καὶ διαπλάσει τὸ πάθημα.

[83] Compare the role of σχήματος ἀλλαγή at *De theriaca ad Pisonem* 14. 252 K, and *De causis contentivis* 3. 4 with *De elementis* 1. 413 K; cases where Galen groups together Empedocles, Democritus, and Epicurus' doctrine on mixture. Someone capable of doing that is capable of anything.

The matter is discussed again in bk. 12 of the *De methodo medendi*.[84] Here Galen notes that there is general agreement amongst adherents of the different medical sects that not all parts of the body are αἰσθητικά. But, he says, if the body is seen to be made up from atoms or ἄναρμα then once again it will be impossible to explain how pain is generated and, ultimately, how any parts of the body can be αἰσθητικά. In a direct reference to Asclepiades' theory[85] he says that when bodies such as the ἄναρμα not only touch, but also break on one another, as is their wont, they will not cause pain since they can neither feel nor be felt—unless, that is, we can say that stones feel pain when they are broken.[86]

Any distinction we might discern in Galen between mutual contact of particles and actual physical breakage could be taken as an implicit distinction between Asclepiades' theory and the theories of Democritus and Epicurus. But what is important is that Galen is nowhere seriously interested in making this kind of distinction. His real enemy in this chapter (and, of course, the whole of the *De methodo medendi*) is the menace of Methodism. The pathology of the Methodists is based, so far as Galen is concerned, on 'the hypothesis of the impassive elements'. However extraordinary a picture of Methodism this might seem to us (let alone most Methodists), Asclepiades was not by any means the only enemy Galen had in this area.[87]

What, then, about Epicurus and Asclepiades? Supporters of

---

[84] *De methodo medendi* 10. 851 K.

[85] Gottschalk (1980), 47, feels that Galen is not referring to Asclepiades 'in so many words', but the use of ἄναρμος and his statement at 10. 853 K to the effect that this subject was treated in more detail in his work '*On The Doctrines of Asclepiades*' leaves little room for doubt.

[86] Note here also the testimony of Rufus of Ephesus (?), *De anatomia*, pp. 184-5 Daremberg: 'According to Erasistratus and Herophilus, nerves are αἰσθητικά; according to Asclepiades this is not at all the case.'

[87] For the development of this particular line of argument against the Methodists, see *De methodo medendi* 10. 841 ff. K, and then pp. 872—where Galen picks up the assault on Methodism again after dealing with Asclepiades. Compare *De naturalibus facultatibus* 2. 27 K, where there is a vague attack on 'sects' who believe matter to be unalterable, unchangeable, and divided up into fine parts, separated by void interstices, and *De methodo medendi* 10. 267-8 K, where Galen argues (mischievously perhaps) that the fundamental Methodist concepts of Thessalus and his followers have their origins in Asclepiades' corpuscular hypothesis.

the thesis that Asclepiades was either an Epicurean, or at least
heavily influenced by Epicurean atomism, invariably appeal to a
chapter in the Galenic treatise *De theriaca ad Pisonem*, entitled
'Refutation of Asclepiades and Epicurus, who deny alteration
and refer the works of nature to the atoms and corpuscles'. The
title has no ancient authority whatever, and does not appear in
the Arabic versions of the text, so we need not worry about it at
all. The chapter itself is too long to reproduce here; in it, Galen
(some have doubted that it is Galen, but I do not) repeats the
claim that Asclepiades merely plagiarized Epicurus' theory. He
makes a direct equation between the atoms of Epicurus and the
corpuscles, void, and the pores. At first sight this would seem to
leave little room for doubt. Yet closer analysis once again reveals
the author's motives. The problem with Asclepiades' theory, it is
said, is that it does not account for the 'alteration' of drugs
within the body. Drugs, the author argues, cannot undergo
alteration in the body, and therefore will not work if they are
made up out of these strange, insensible, and unalterable corpuscles.

εἰ μὲν γὰρ ἐξ ἀτόμων καὶ τοῦ κενοῦ κατὰ τὸν Ἐπικούρου τε καὶ
Δημοκρίτου λόγον συνειστήκει τὰ πάντα, ἢ ἔκ τινων ὄγκων καὶ πόρων
κατὰ τὸν ἰατρὸν Ἀσκληπιάδην· καὶ γὰρ οὕτως ἀλλάξας τὰ ὀνόματα
μόνον καὶ ἀντὶ μὲν τῶν ἀτόμων τοὺς ὄγκους, ἀντὶ δὲ τοῦ κενοῦ τοὺς
πόρους λέγων τὴν αὐτὴν ἐκείνοις τῶν ὄντων οὐσίαν εἶναι βουλόμενος·
εἰκότως ἂν ἔμενεν ἀναλλοίωτα τὰ φάρμακα, κατὰ μηδὲν τρέπεσθαι
μηδ᾽ ὅλως ἐξίστασθαι τῆς αὐτῶν ποιότητος δυνάμενα. ἐπεὶ δ᾽ οὐκ ἔστιν
ἀληθὴς ὁ λόγος οὗτος, ὡς δείξομεν, ἀλλ᾽ ἀλλοιοῦται, ὡς ἔφην, τὰ
πάντα καὶ τρέπεται ῥᾳδίως καὶ εἰς ἄλληλα τὴν κρᾶσιν λαμβάνει,
ἀνάγκη τῆς κράσεως δι᾽ ὅλων τῶν κιρναμένων γιγνομένης τὸ
ἰσχυρότερον τοῦ ἥττονος κρατεῖν, καὶ διὰ τοῦτο ἡμεῖς ταῖς ἐντέχνοις
μίξεσι πρὸς τὴν χρείαν τῆς ἐνεργείας τὰς ποιότητας τῶν φαρμάκων
ἐναλλάσσομεν, οὐκ ἂν δυναμένου τούτου γενέσθαι, εἰ μικρά τινα ἦν
καὶ ἀπαθῆ καὶ ἄτρεπτα τῆς οὐσίας τὰ σώματα.

Is the universe made up of atoms and void, as the theory of Epicurus
and Democritus would have it, or from some kind of 'corpuscles' and
'pores', as Asclepiades the doctor claims? (Asclepiades simply changed
the names, talking about 'corpuscles' instead of atoms, and 'pores'

instead of void, but still subscribed to the same concept of the nature of matter.) If this were the case, then it would be reasonable to imagine that drugs remain unchanged, are completely unalterable, and do not submit to qualitative change. I will demonstrate the falsity of this argument. Drugs do change, as I have said; they all readily undergo alteration, and admit of mixture with one another. The stronger overcomes the weaker when their substances are thoroughly blended. Consequently, we alter the qualities of drugs using certain techniques of combination, with a view to the utility of their action. If tiny, impassive, and unalterable particles of matter existed, then this would not be possible.[88]

Once again Galen lets his secret out. He effectively says that ἀπαθής is equivalent in sense to ἄτρεπτος. The particles of Asclepiades are unchangeable, and cannot be altered in terms of their own qualities, and this means that what is made out of them cannot be altered either. We are then treated to some examples of what the author means by 'change' in this context: when you put your finger into cold water, the whole body changes its state. The author had a friend who could tell what kind of wind was blowing by feeling it; women abort when they hear thunder.[89] The argument is summed up in these terms—is it not absurd to explain qualitative alteration by appealing to its negation? Throughout this chapter in the *De theriaca ad Pisonem* it has suited the author's purpose to assimilate Asclepiades to a traditional form of Abderite atomism in order to make him sound more ridiculous. The catch-all remark at 14. 252–3 K—ἡ γὰρ ποιὰ τῶν ὄγκων μετατιθεμένων σύνθεσις τοῦ μὲν σχήματος ἀλλαγὴν μόνην ἐργάζεται, ἀλλοίωσιν δὲ καὶ ποιότητα ἄλλην ἐξ ἄλλης γεννῆσαι ἀδυνατεῖ, 'the nature of the compound which results from the composition of the corpuscles produces only a change in arrangement; it is impossible for alteration to take place, or for one quality to come from another'[90]—could apply to nearly all ancient monistic theories, atomistic or not.

---

[88] 14. 250–1 K.
[89] Ibid. 252 K.
[90] There are more examples of this type of argument in Galen: see in particular *De naturalibus facultatibus* 2. 101–18 K; *De causis contentivis* 3 ff., 5 ff.

Galen was no fool. He must have known that Democritean and Epicurean atomism differed profoundly from Asclepiades' corpuscular hypothesis. After all, he was aware of the fragility of the particles. And in his note at *De elementis* 1. 418 K he makes it quite clear that he knows that the Democritean atoms are unbreakable 'on account of their hardness' and 'indivisible on account of their size'. He is just as clear about the Asclepiadean corpuscles. Granted, θραυστός occurs only once in surviving Greek texts of Galen,[91] but this serves only to strengthen our suspicions that ἄναρμος (far more common) carried at least some of these connotations of breakability. We may infer that Galen also saw the θραύσματα as being ontologically identical to the particles from whose collisions they arose. Certainly, there is no clear evidence in Galen of any distinction between different levels of particle in Asclepiades' theory. Conceptions of what 'quality', 'activity', and 'passivity' might mean are informed for Galen much more by the concerns of his own pathology along with a rather extreme form of continuum theory than by immersion in earlier philosophical discussions of related problems. However much Galen knew about philosophy—and he knew a great deal—he knew, and cared more about, the Hippocratics.

All this also militates strongly against taking the labels which Galen applies to the particles in different corpuscular theories as equivalent in the way that Gottschalk does. Surely it is going a little far to suggest, as he does, that Galen really believes that ἐλάχιστα, ἄναρμα, and μέρη are roughly equivalent terms.[92] That Galen is quite happy to be vague, for reasons which have little to do with preserving the integrity of the doxographies, and a lot to do with rhetoric, seems to me clear from passages such as this one from the *De sanitate tuenda*:

---

[91] At *De constitutione artis medicae* 1. 249 K.

[92] Gottschalk (1980), 41. For cases where Galen would have us believe that the terms are roughly similar, see *De elementis* 1. 416–17 K, *De usu partium* 4. 350 K; *Commentarium in Hippocratis de natura hominis* 15. 36 K. (At *De dignoscendis pulsibus* 8. 927 K, ἄναρμα is effectively glossed with ἄτμητα (uncuttable): σώματα σμικρὰ καὶ ἀδιαίρετα καὶ ἄτμητα καὶ ἄναρμα.

συμμετρία γὰρ δή τις ἡ ὑγίεια κατὰ πάσας ἐστὶ τὰς αἱρέσεις, ἀλλὰ
καθ᾽ ἡμᾶς μὲν ὑγροῦ καὶ ξηροῦ, καὶ θερμοῦ καὶ ψυχροῦ, κατ᾽ ἄλλους
δὲ ὄγκων καὶ πόρων, κατ᾽ ἄλλους δὲ ἀτόμων, ἢ ἀνάρμων ἢ ἀμερῶν ἢ
ὁμοιομερῶν . . .

Health is some kind of balance for all the sects; for us it is a balance of
wet and dry, hot and cold, but for others a balance of corpuscles and
pores, for others of atoms, or ἄναρμα, or partless bodies, or homoiomeries.[93]

Vagueness apart, though, it is certainly most interesting to see
that Galen should refer so often to this *group* of theories when
he attacks Asclepiades. As far as I know, there is no evidence
that any other corpuscular hypotheses were adopted by doctors.
In particular, there is no evidence for an 'Epicurean' school of
medicine.[94] And Galen tends to reserve this kind of spleen for
professional rivals. In the next chapters, we shall find out more
about who they might have been.

There need be no contradiction, then, between Sextus and
Galen over the quality of the corpuscles. We merely need to be
clear about what the 'qualities' in question actually are. There is,
however, still an unresolved problem lurking in the Caelius text
with which I began this discussion. It seems most natural that,
regardless of what is done with *secunda* at *De morbis acutis* 1. 105,
the clause beginning 'corpuscula . . . moventia' is to be taken as a
gloss, or perhaps depends parenthetically upon the first part of
the sentence. Only one type of particle is being referred to here.
The particles are 'sine ulla qualitate solita', not 'sine ulla
qualitate'. At *De morbis acutis* 1. 113, Caelius tells us that the
qualities the very finest particles lack are those of being hot or
cold, or of being able to submit to tactile sensation. Without
recourse to emendation,[95] we might explain this problem by
invoking the same kind of ambiguity around the use of *qualitas*
in pathological contexts as we have noted in Galen. Whatever

[93] *De sanitate tuenda* 6. 15 K. Sextus does much the same at *Adversus mathematicos* 10. 254.

[94] As opposed to Epicurean interest in medical matters, for which there is some evidence: see Epicurus, fr. 18 Arrighetti, Philodemus, *De signis* 36. 1, with Gigante (1975).

[95] Although I have been sorely tempted to read 'solida' for 'solita'.

the case, it is quite clear that Caelius is not using *qualitas* in a
sense equivalent to Sextus' use of ποιά, even if his Methodism
precluded its use in a Galenic manner.

We are now in a position to suggest that the *corpuscula* and
fragmenta of Caelius, the ὄγκοι of Galen, the *moles* of Calcidius,
and the ἄναρμοι ὄγκοι of Sextus are all the same. Why does
Caelius call the *corpuscula primordia*, when the *fragmenta* referred
to in the next sentence would seem to fit that description rather
better? Because they are all the same. Asclepiades' corpuscles
are *primordia* in the sense that they are the ultimate constituents
of matter, but they are always *secunda* because they are fragile. It
is possible that editors athetized *secunda* because of its apparent
contradiction with *primordia*, and because of the fact that it takes
a syntactic act of faith to see it as introducing a second, contrast-
ing clause. In as much as there is one at all, the distinction
between fragments and corpuscles is informed by the dynamics
of Asclepiades' theory, and by his views on bodily processes such
as digestion and respiration. Our witnesses applied a variety of
different epithets to the particles; some, like Sextus, because
they were looking at Asclepiades in a philosophical rather than
medical tradition, and others, like Galen, because they quite
simply had no interest in presenting the theory for its own sake.
When Galen says, at *De naturalibus facultatibus* 2. 39 K, that
'Asclepiades ... divided and broke up matter into ἄναρμα
elements and ridiculous ὄγκοι', he is not suggesting that there
were two distinct types of particle. He is just being rude.

So the corpuscles *were* divisible after all. But, as I asked at the
outset, how far? When Caelius says that they shatter into 'infinite fragments' he is, whether he knows it or not, distancing
Asclepiades quite a long way from the atomistic tradition. In saying that the parts of the body are corpuscles which can be
divided into innumerable further parts, Caelius is implying that
the Asclepiadean body had, potentially at least, an infinite
number of parts. This is anathema to Epicurean philosophy.[96]

---

[96] At *Ad Herodotum* 56–7 (a passage interesting for its use of the word ὄγκοι) Epicurus
argues thus: 'we should not assume that in a limited body there exists a limitless number
of ὄγκοι, whatever their size.' Lucretius takes up the same point at *De rerum natura* 1.

Final judgement on this will have to wait until we have considered the role of the pores. We may for the moment, however,
conclude that there is a strong possibility that Asclepiades'
particles were infinitely divisible.

Far from being atoms, then, as Galen and some others would
have us believe, the corpuscles seem to have been described
precisely in order to contrast them with the impassive atoms of
Democritus and Epicurus. We have little choice but to entertain
the possibility that Galen has been radically tendentious on this
whole issue. Ultimately we will have to ask why: not only why
Galen did this, but why Asclepiades chose to adopt such an
extraordinary theory. This is still some way off. The next task is
to consider the Asclepiadean doctrine on void.

615–16 in his discussion of minimal parts. (Without a limit, *corpora constabunt ex partibus infinitis*.)

2
Void?

I have suggested that the various terms *corpuscula*, ἄναρμοι ὄγκοι, and so on all refer to the same fragile particles. I argued that the fragility of Asclepiades' corpuscles was something which set them apart from the indivisible atoms of Democritus and Epicurus. This is a telling blow against those who believe that Asclepiades took over the Epicurean theory. On my argument Galen has shown himself either perverse or up to something more than a little strange when he stresses the connection between Asclepiades and Epicurus so vigorously. I shall approach Galen in this chapter with rather more caution. But what of all those cases where the *pores* in Asclepiades' theory are explicitly equated with the void of Epicurus?

There are one or two implicit cases of this kind of identification which could conveniently be dismissed as doxographical simplification. Dionysius of Alexandria in his work *On Nature* (quoted by Eusebius) summarized ancient physical theories. He was not really interested in *any* differences between the 'Monistic' theories because they were anti-teleological and therefore anti-Christian. The main butt of Dionysius' attack was Epicurus, the arch-advocate of a Nature without purpose, and all those who held even vaguely similar theories were assembled around him:

πότερον ἕν ἐστι συναφὲς τὸ πᾶν, ὡς ἡμῖν τε καὶ τοῖς σοφωτάτοις Ἑλλήνων Πλάτωνι καὶ Πυθαγόρᾳ καὶ τοῖς ἀπὸ τῆς στοᾶς καὶ Ἡρακλείτῳ φαίνεται, ἢ δύο, ὡς ἴσως τις ὑπέλαβεν, ἢ καὶ πολλὰ καὶ ἄπειρα, ὥς τισιν ἄλλοις ἔδοξεν, οἳ πολλαῖς τῆς διανοίας παραφοραῖς καὶ ποικίλαις προφοραῖς ὀνομάτων τὴν τῶν ὅλων ἐπεχείρησαν κατακερματίζειν οὐσίαν, ἄπειρόν τε καὶ ἀγέννητον καὶ ἀπρονόητον

ὑποτίθενται; οἱ μὲν γὰρ ἀτόμους προσειπόντες ἄφθαρτά τινα καὶ
σμικρότατα σώματα πλῆθος ἀνάριθμα, καί τι χωρίον κενὸν μέγεθος
ἀπεριόριστον προβαλόμενοι, ταύτας δή φασι τὰς ἀτόμους ὡς ἔτυχεν ἐν
τῷ κενῷ φερομένας αὐτομάτως τε συμπιπτούσας ἀλλήλαις διὰ ρύμην
ἄτακτον καὶ συμπλεκομένας διὰ τὸ πολυσχήμονας οὔσας ἀλλήλων
ἐπιλαμβάνεσθαι, καὶ οὕτω τόν τε κόσμον καὶ τὰ ἐν αὐτῷ, μᾶλλον δὲ
κόσμους ἀπείρους ἀποτελεῖν. ταύτης δὲ τῆς δόξης Ἐπίκουρος γεγό-
νασι καὶ Δημόκριτος. τοσοῦτον δὲ διεφώνησαν ὅσον ὁ μὲν ἐλαχίστας
πάσας καὶ διὰ τοῦτο ἀνεπαισθήτους, ὁ δὲ καὶ μεγίστας εἶναί τινας ἀτόμους ὁ Δημόκριτος ὑπέλαβεν. ἀτόμους δὲ εἶναί φασιν ἀμφότεροι καὶ
λέγεσθαι διὰ τὴν ἄλυτον στερρότητα. οἱ δὲ τὰς ἀτόμους μετο-
νομάσαντες ἀμερῆ φασιν εἶναι σώματα, τοῦ παντὸς μέρη, ἐξ ὧν
ἀδιαιρέτων ὄντων συντίθεται τὰ πάντα καὶ εἰς ἃ διαλύεται. καὶ τούτων
φασὶ τῶν ἀμερῶν ὀνοματοποιὸν Διόδωρον γεγονέναι. ὄνομα δέ,
φασιν, αὐτοῖς ἄλλο Ἡρακλείδης θέμενος ἐκάλεσεν ὄγκους, παρ' οὗ
καὶ Ἀσκληπιάδης ὁ ἰατρὸς ἐκληρονόμησε τὸ ὄνομα.

Is the universe a unity, as we believe, along with Plato and Pythagoras,
the wisest of the Greeks, and the Stoics and Heraclitus. Or is it a dual-
ity, as some might perhaps imagine? Or is it composed of an infinite
number of parts? This was the belief of those people who trusted their
own undisciplined minds and tried to cut up the existence of the
universe under cover of fancy terminology. They held the universe to
be infinite, ungenerated, and unpremeditated. Some of them posited
atoms—imperishable, very tiny bodies, infinite in number. They
imagined the existence of void space, infinite in extent. They claimed
that these atoms are borne about of their own accord in the void; that
they collide with one another in their disorderly rush and are woven
together because they have a variety of shapes. This is how they
explain the generation of the cosmos and its contents—or rather the
generation of innumerable cosmoses. Epicurus held this view, and so
did Democritus. But they differed from each other to this extent—
Epicurus held all atoms to be *minima*, and for this reason impercept-
ible, while Democritus held that some atoms could be very large. But
both of them said that they are atoms (in the sense that they are indivisible) because of their hardness, and that they cannot be dissolved.
On the other hand some people changed the name 'atoms' and called
them 'partless bodies', parts of the whole, and argued that out of these
indivisible entities all things are assembled, and into them all things
are dissolved. Apparently Diodorus coined the term 'partless bodies'. It

is said that Heraclides gave them another name—ὄγκοι—and that Asclepiades the doctor took the name from him.[1]

In another passage, pseudo-Hero of Alexandria draws an analogy between the atoms and void of Democritus, and the corpuscles and pores of Asclepiades; this forms part of his analysis of the meaning of the term 'hypothesis'.[2] This text seems to spring from the same source as Sextus Empiricus' discussion of the word at *Adversus mathematicos* 3. 5. Sextus does not mention Democritus at all, but goes into the basic premisses of Asclepiades' theory in unusual detail. We might reasonably explain away this Heronic testimonium as showing signs of Galenic infection. More simply still we could dismiss both Hero and Dionysius: this might be best. Dionysius, after all, says that Diodorus Cronus merely took over Abderite atomism, substituting his own term, *minima*, for previous ones. Diodorus clearly did more than that. And as for the notion that Democritean atoms could be very large—I need say nothing about that. But Galen himself is adamant, and he cannot be explained away so easily. He insists that Asclepiades held a conception of void which was identical to that of Epicurus in all but name. Sometimes he talks about ὄγκοι καὶ κενόν ('corpuscles and void').[3] Sometimes he goes further: τὸ δὲ κενὰς εἶναί τινας χώρας ἢ κατὰ τὸ ὕδωρ ἢ κατὰ τὸν ἀέρα, τῇ μὲν Ἐπικούρου τε καὶ Ἀσκληπιάδου δόξῃ περὶ τῶν στοιχείων ἀκόλουθόν ἐστι, 'It follows upon the Epicurean and Asclepiadean doctrine on the elements that there exist some kind of void spaces, whether in water or air'.[4] He makes a variety of asides like this—Ἐπίκουρος μὲν οὖν, καίτοι παραπλησίοις Ἀσκληπιάδῃ στοιχείοις εἰς τὴν φυσιολογίαν χρώμενος, 'Epicurus, although he employs elements in his physics similar to those of Asclepiades'.[5] But nowhere does he

[1] Eusebius, *Praeparatio evangelica* 14. 23.

[2] Ps.-Hero, *Definitiones*, p. 166. 1–15 Heiberg.   [3] e.g. *De usu partium* 3. 474 K.

[4] *Commentarium VI in Hippocratis librum Epidemiarum VI* 17в. 162 K.

[5] *De naturalibus facultatibus* 2. 45 K. Cf. *De simplicium medicamentorum temperamentis ac facultatibus* 11. 405 K: λέγω δὴ ἀραιὰν ἧς τὰ μόρια διαλαμβάνεται χώραις κεναῖς, ἐπισταμένων ἡμῶν δηλονότι καὶ μεμνημένων ἀεὶ πῶς λέγεται χώρα κενὴ πρὸς τῶν ἡνῶσθαι φασκόντων τὴν οὐσίαν, ὅτι μὴ καθάπερ Ἐπικούρῳ καὶ Ἀσκληπιάδῃ δοκεῖ, ἀλλ' ἐστιν ἀέρος πλήρης ἐν ἅπασι τοῖς ἀραιοῖς σώμασιν ἡ κενὴ χώρα; and *De theriaca ad Pisonem* 14. 250 K, which I discussed in ch. 1.

explain the similarity in any detail, and nowhere does he give any clear hint about the role which void played in Asclepiades' physiology. The ἄναρμοι ὄγκοι are all very well, but what about the pores? Do they correspond to Epicurean void?

This aspect of the problem has received remarkably little attention. In the best modern account, Harig[6] argues that Asclepiades was indeed a void theorist; on his argument Asclepiades took the ἄναρμοι ὄγκοι of Heraclides of Pontus and added the refinement of an Epicurean void so as to give the particles room in which to move. This would certainly save Galen's evidence. But is that enough? Surely Heraclides had thought of the problem of explaining motion. (It was not as if the physicists of the Academy or Lyceum had not thought about it.) The fragility of the Asclepiadean corpuscles points to a concept of matter (and void) far removed from the atomists. It smacks of the possibility of infinite divisibility, anathema to physicists who needed to cling on to a strict antithesis between corporeality and incorporeality.

By this time in antiquity void came in many guises, often highly sophisticated. In fact, to call them all 'void' is somewhat misleading, especially given the degree of involvement of prominent anti-void, or continuum, theorists in the debate. Philosophers and scientists had gone a long way from the concepts of absolute 'being' and 'not being' which the fifth-century atomists had been trying to wrestle out of the hands of the Eleatics. Amongst the atomists, Epicurus had insisted that void could be understood in general, almost all-embracing, terms as some kind of intangible 'substance', a three-dimensional extension which can happily exist independently of body as well as providing body with a spatial substrate.[7] This kind of void is physically passive and has no power, but it gives the atoms their crucial room to move. Epicurus positively affirmed its existence; it was not a shady concept, overshadowed by matter.

---

[6] Harig (1983).
[7] The ἀναφὴς φύσις: *Ad Herodotum* 39–40; Aetius, *Placita* I. 20. 2 = Long and Sedley (1987), 5 C. On the Epicurean characterization of void see e.g. Sedley (1982). This brief account of void is based on Simplicius' version of Proclus' account at *Corollarium de loco*, p. 601, 14–24.

For Stoics like Chrysippus, talking loosely about void was not good enough. Void within the world could not be separated from the actual existence of body. Void on these terms was characterized as 'place' or 'room', while void in the sense of a three-dimensional extension *actually* empty of body could exist only outside the world.[8] These terminological distinctions were important.

Some theorists were even more severe. 'Place' for Plato was absolutely without extension beyond the bodies whose material substrate it provided. Others took a more lenient view. Strato of Lampsacus seems to have developed a theory in opposition to Aristotle that void might exist within the cosmos but only as a three-dimensionally extended interval between a body and what contains it. That is to say, void could exist in the gaps between ill-fitting particles of matter, while void space on any larger scale, not existing in such interstices, could only be created by force.[9] With this theory, void came of age; it became vacuum and assumed a power of its own.

Alongside the variety of concepts of void on offer, there was also a tradition of particulate theories of matter which were not strictly atomistic at all. The different corpuscular models of matter attributed to Anaxagoras, Empedocles, Plato, Xenocrates, Strato, Heraclides of Pontus, and Diodorus Cronus, let alone Erasistratus (among the doctors), seem to have offered no place to void in the Democritean or Epicurean sense.[10] (Even though indivisible magnitudes seem to figure in the theory of Xenocrates (a Platonist).) Quite apart from the suggestion that Asclepiades merely took over the system of Heraclides, there are tantalizing terminological parallels between the language of Asclepiades' corpuscular hypothesis and those of Empedocles and Xenocrates.[11]

[8] Long and Sedley (1987), 49 A-J.
[9] See Gottschalk (1965), 127-41; Furley (1989).
[10] Even though the doxographers group many of them together: e.g. Aetius, *Placita* 1. 24. 2 (=31 A 44 DK).
[11] For the testimonia associating Empedocles and Xenocrates with a theory of *minima*, see Aristotle, *De generatione et corruptione* 334ᵃ26-30. Aetius, *Placita* 1. 13. 1, calls the Empedoclean *minima* θραύσματα ἐλάχιστα; cf. 1. 17. 3. Galen at *Commentarium in Hippocratis librum de natura hominis* 15. 49-50 K attributes a belief in μικρὰ μόρια to Empedocles.

It is only sensible, then, that we should embark with an open mind on an examination of the Asclepiadean pores, and his dynamic principle πρὸς τὸ λεπτομερὲς φορά—'movement towards what is fine', hereafter called PTLP).

Theories which have room for pores or gaps in or around solid bodies can be at home in either 'atomistic' or continuum systems. The term πόρος is indifferent between them. I have already noted some medical connotations of the term ὄγκος, and suggested that these should not be ignored in favour of more 'philosophical' senses when we look at Asclepiades' motivations and intellectual affiliations. The case with πόρος is similar. In an original sense of 'ford' it stands for something (in this case full of water) which allows objects to pass through it. Its use was widespread in early philosophy and medicine, and it referred generally to any type of passage which permitted communication between different bodies.[12] All this is hardly surprising, and Lonie put it rather well: 'In a sense, all Greek medical theories are about πόροι; the human body is simply a system of πόροι.'[13] Indeed, terms such as πόρος, ἀγγεῖον, ὄχετος, and ὁδός abound in the Hippocratic Corpus, and the student of Asclepiades should not get too excited when he finds them in the Asclepiadean testimonia.[14] Such words are often, though not always,[15] used to refer to visible passages which can be discovered anatomically. Aristotle uses πόρος (usually qualified in some way) of blood-vessels and the urethra; the term has a history, perhaps going back to Alcmaeon, of being applied to various sensory tracts—in particular the optic 'nerve'.[16] Galen uses the word in a variety of ways, some of which look remarkably close to those he attributes to Asclepiades. But most often Galen's pores are visible passages: the urethra (or ureter), for

---

[12] Guthrie (1965), 234 n. 3; Wright (1981), 230.
[13] Lonie (1965), 128.
[14] Lloyd (1983), 153 n. 116; Duminil (1983), pt. 1.
[15] For a counter-example, see *On Regimen* 1. 6. 514 Littré.
[16] e.g. Aristotle, *De generatione animalium* 743ᵃ8, 782ᵇ1; the urethra: *Historia animalium* 493ᵇ4. A fragment of Aristotle in Athenaeus (*Deipnosophistae* 1, 24ᴇ=fr. 236 Rose) compares the skin to a colander in an explanation of perspiration. For Alcmaeon see Lloyd (1975); on the extensive use of the term by Theophrastus see Gottschalk (1961).

instance,[17] the 'bile-ducts',[18] the windpipe,[19] the 'spermatic ducts',[20] the ethmoid passages,[21] the 'optic passage',[22] and the 'acoustic passage'.[23] Galen is always keen to set himself up as someone who does not believe in what he cannot see by experiment and observation, and he frequently qualifies πόρος with αἰσθητός.[24] Elsewhere,[25] he discusses the morbid effects of macroscopic blockages in macroscopic pores, ever careful to distinguish this position from that of the Methodists, who, he argues, believe that phenomena such as these exist at the theoretical level. But even Galen is human. He believes in 'tiny pores in the skin and throughout the whole body' (in contexts which have nothing to do with Asclepiades) which form part of his explanations of processes such as perspiration.[26] One thing, however, is quite clear: invisible or not, Galen saw no similarity between his own pores and those of Asclepiades.

Empedocles, among those of the Presocratics with supposed medical connections, is particularly associated with the early use of a theory of pores to explain a wide variety of natural phenomena ranging from respiration to sense-perception and what we now call magnetism. Empedoclean pores were not empty in the atomistic sense. A considerable amount of work has been done

---

[17] *De constitutione artis medicae* 1. 282 K; *De naturalibus facultatibus* 2. 31 K; *De usu partium* 3. 377 K; *De locis affectis* 8. 18, 393, 394, 404 K; *De methodo medendi* 10. 527 K.

[18] *Ars medica* 1. 354 K; *De temperamentis* 1. 631–2 K; *De naturalibus facultatibus* 2. 40–1 (this last in connection with Asclepiades, but the use is Galen's), 67, 93 K; *De usu partium* 3. 301, 305 K; *De sanitate tuenda* 6. 74 K; *De methodo medendi* 10. 908 K; *Commentarium in Hippocratis Aphorismos* 17B. 745 K.

[19] *De usu partium* 3. 443 K.

[20] *De naturalibus facultatibus* 2. 35 K.

[21] Ibid. 58 K; *De instrumento odoratus* 2. 881–3 K (the olfactory passages); *De locis affectis* 8. 214 K.

[22] *De nervorum dissectione* 2. 833 K; *De usu partium* 3. 813 K; *De causis symptomatum* 7. 89 K (on Herophilean usage).

[23] *De ossibus ad tirones* 2. 745 K; *De causis symptomatum* 7. 103 K; *De locis affectis* 8. 21–2, 234 K; *De methodo medendi* 10. 352 K; *De compositione medicamentorum secundum locos* 12. 599 K; *Commentarium VI in Hippocratis librum VI Epidemiarum* 17B. 267 K etc.

[24] See esp. *De tumoribus praeter naturam* 7. 715 K.

[25] *De differentiis morborum* 6. 856–62 K; and cf. *De praesagitione ex pulsibus* 9. 384–5 K.

[26] *Ars medica* 1. 393 K; *De inaequali intemperie* 7. 736 K; *De difficultate respirationis* 7. 940 K; *De methodo medendi* 10. 499 K; *De simplicium medicamentorum temperamentis ac facultatibus* 11. 402 K. At *De sanitate tuenda* 6. 66 K Galen speaks of 'tiny pores which fill the whole body'; cf. 6. 218 K, although it is not clear there that the doctrine of ἔμφραξις is Galen's own.

on Empedocles' impact on Hippocratic medicine,[27] and although much of it is rather conjectural, at least one ancient medical sect claimed him as an ancestor.[28] The truth of that claim is quite another matter; there is no clear evidence to suggest that there was any direct Empedoclean influence, say, on Asclepiades via the 'Sicilian connection'. But it does seem likely that the influence of Empedocles on later medical theories has been seriously underestimated.

If we accept that the term πόρος does not help us diagnose the presence or absence of void and that several competing characterizations of void were on the philosophical and scientific market, this does not necessarily mean that the distinction between atomism and continuum theory became particularly blurred, even after Strato. Galen is one of the most hardline continuum theorists of them all. So the mere fact of Asclepiades' having used the terminology of pores is unremarkable. We shall have to press on further. Is there anything in the Asclepiadean theory which necessitates the supposition that these pores in effect contained void?

One key explanatory demand which illuminates the philosophical development of the concept of void concerns motion. The need to account for movement, and consequently change, is obvious. The atomists needed to have some way of explaining how solid objects move through other solid objects. The solution was plain: objects are not solid. Epicurus on occasion used the idea of pores to describe the interstices in matter compounded out of the absolutely solid atoms. It was through these interstices that other bodies could pass.[29] But the problem of motion in a continuum had been dealt with by (among others) Empedocles, Plato, and Aristotle through the theory of *antiperistasis*, or 'reciprocal replacement'. This could account for motion through areas of matter which were not so much porous in the

---

[27] See Jouanna (1961); Longrigg (1976), 436–8; and esp. Mourelatos (1987), 166–91. However problematic the question of Empedoclean influence, it seems less so than that of Democritean influence, *pace* Wellmann (1929).

[28] Pliny, *Naturalis historia* 29. 5–6 and ps.-Galen, *Introductio sive medicus* 14. 683–4 K.

[29] *Ad Herodotum* 47, 61; *Ad Pythoclem* 107, 110–11; frr. 24. 46. 13 ff.; 31. 4. 1–4; 35. 10. 1–15 Arrighetti.

vacuous sense, as less dense than what surrounds them. One object can yield its place to another in a continuum without any void being present. Such a theory is particularly associated with Theophrastus' explanations of how fire interacts with its fuels.[30] Even a fish can move in water, because it can force the water in its path up and around its body, so as to occupy the space the fish occupied a moment before. (The impossibility of this explanation was a key Epicurean argument *in favour* of void.) The Anonymus Londinensis attributes to Asclepiades the statement that 'body cannot pass through body', but even though this phrase looks as though it might be atomistic it does not necessarily tell us much about Asclepiades' true affiliations.[31]

Galen is as good a representative of the continuum theorists as any. When he speaks about 'what is empty' ($\kappa\epsilon\nu\acute{o}\nu$), he takes care to avoid misinterpretation by his readers. In the *De simplicium medicamentorum temperamentis ac facultatibus* he leaves no room for doubt about what he means by his concept of relative density. What is light ($\dot{\alpha}\rho\alpha\hat{\iota}os$), he says, contains many 'empty spaces' ($\chi\hat{\omega}\rho\alpha\iota\ \kappa\epsilon\nu\alpha\acute{\iota}$), but by 'empty' he means 'full of air'. There is air in all rarefied bodies.[32] There is little that is surprising about this; the most hardened continuum theorist can still have an empty stomach or be faced with an empty bottle without having his whole theory collapse around him. Yet Galen is the chief source for our knowledge of the nature of Asclepiades' pores, and we should be aware how easy it can be to confuse Galenic and Asclepiadean terminology. Galen himself argues a little further on in this same passage that dry reeds burn well 'because of the number of pores in them which were formerly filled with air', while still leaving room to attack Asclepiades and Epicurus in another treatise for holding that there are empty spaces (again, $\chi\hat{\omega}\rho\alpha\iota\ \kappa\epsilon\nu\alpha\acute{\iota}$) in water and air which permit the

[30] e.g. *De igne* 28, 38, 42, 45, 61; and Gottschalk (1961).
[31] Anonymus Londinensis 39. 1–4; the phrase reappears, attributed to Asclepiades, in the Yale Scholiast to Galen's *De elementis* (=3. 245 in Moraux (1977)), where it seems to be set up as an argument against the Stoic theory of 'complete mixture'.
[32] 11. 405 K: see above, n. 5.

propagation of light and sound.[33] Even here there is nothing about the position of Asclepiades which makes it essential for us to agree with Galen and adopt him as an Epicurean void theorist. The example of the propagation of light in water looks remarkably similar to Hero of Alexandria's explanation of the passage of light through πόροι in the medium through which it is transmitted.[34] Hero was no Epicurean, even if he accepted some ideas which Epicurus might conceivably have shared. No more need this be the case with Asclepiades. In fact it is generally thought that Hero is following Strato's theory quite closely here. There are bound to be non-Epicurean treatments of the problem which superficially at least seem to share common ground with Epicurean doctrines. But the Hero/Strato theory seems to have involved the assertion that void within the world *only* exists in the interstices between different particles of matter, and that void on any scale larger than this can only be produced by force. A shared illustration does not make for a shared theory.

There are at least three preliminary answers to the problem of Asclepiades' pores. They may have been continuous (and therefore Galen is lying, or wrong); they may have represented or contained Epicurean void (in which case the theory of the ἄναρμοι ὄγκοι becomes still stranger); or they might in some way be related to the small-scale interstitial void of Strato. It is time to consider the pores in action.

The feature of Asclepiades' pores which comes in for most criticism in Galen is the fact that they were 'theoretical' (ἐν λόγῳ θεωρητόν), like the corpuscles. In the *De naturalibus facultatibus* especially, Galen sees in these pores which exist simply because Asclepiades decided that they did, an example of the kind of thinking which led Asclepiades away from the proper observation of the phenomena of disease, away from

---

[33] *Commentarium VI in Hippocratis librum VI Epidemiarum* 17B. 161-2 K. Cf. also *De dignoscendis pulsibus* 8. 927-9 K, where Galen attacks those 'who break up the continuum of the universe' by void: all such doctrines, he says, involve the positing of small, indivisible, and ἄναρμα bodies. Also *De naturalibus facultatibus* 2. 27-8 K, and esp. *De differentiis pulsuum* 8. 670-85 K.

[34] Hero of Alexandria, *Pneumatica*, p. 26 Schmidt.

anatomy and, most seriously of all, from teleology. The difference between Galenic and Asclepiadean pores is neatly illustrated by the attack on Asclepiades' explanation of how urine gets into the bladder. Galen believes that urine passes through two visible pores (the ureters), while he tells us that Asclepiades had a more complicated explanation:

βούλεται γὰρ [sc. ὁ Ἀσκληπιάδης] εἰς ἀτμοὺς ἀναλυόμενον τὸ πινόμενον ὑγρὸν εἰς τὴν κύστιν διαδίδοσθαι κἄπειτ' ἐξ ἐκείνων αὖθις ἀλλήλοις συνιόντων οὕτως ἀπολαμβάνειν αὐτὸ τὴν ἀρχαίαν ἰδέαν καὶ γίγνεσθαι πάλιν ὑγρὸν ἐξ ἀτμῶν ἀτεχνῶς ὡς περὶ σπογγιᾶς τινος ἢ ἐρίου τῆς κύστεως διανοούμενος, ἀλλ' οὐ σώματος ἀκριβῶς πυκνοῦ καὶ στεγανοῦ δύο χιτῶνας ἰσχυροτάτους κεκτημένου.

Asclepiades would have it that the liquid we drink is dissolved into vapours and that these pass into the bladder. The vapours then recondense and resume their original form, becoming water once more from vapour. In effect Asclepiades sees the bladder as a sort of sponge, or piece of wool, and not the entirely solid and watertight body it is, possessed of two very tough coats.[35]

What ends up as urine is first broken down into vapour before passing into the bladder through countless tiny pores in its coat. (We can imagine that the process involves the dissolution of large corpuscles which make up the urine into fine ones which constitute the vapour. Galen uses the verb ἀναλύω of this process of dissolution. It is cognate with *solvo*.) The Asclepiadean bladder is 'porous';[36] Galen asks why, if urine can find its way like this into the bladder, it does not vaporize and travel throughout the whole body. The point seems a good one; Soranus records that the Asclepiadean body was so porous that an ointment applied to one part will soon be distributed everywhere.[37]

There is only one surviving discussion of the constitution of

---

[35] *De naturalibus facultatibus* 2. 32 K.

[36] Galen illustrates the Methodist concept of pores with an analogy of an earthenware pot at *De sectis ad introducendos* 1. 100 K.

[37] Soranus, *Gynaecia* 1. 9. 35 ὁ γοῦν Ἀσκληπιάδης φησιν ἐπιτιθεμένης τῆς διὰ πηγάνου κηρωτῆς ἑλκωμένῳ τὰ σκέλη ἀντίληψιν τῷ κάμνοντι γενέσθαι τῆς ποιότητος κατὰ διάδοσιν.

the pores. It comes in the passage from Caelius Aurelianus which was the centre-piece of Chapter 1. I continue my quotation from *De morbis acutis* 1. 106:

fieri etiam vias ex complexione corpusculorum intellectu sensas, magnitudine atque schemate differentes, per quas sucorum ductus solito meatu percurrens, si nullo fuerit impedimento retentus, sanitas maneat, impeditus vero statione corpusculorum morbos efficiat.

Theoretical pores come to be out of the interweaving of the corpuscles. The pores differ in size and shape, and it is through them that the passage of bodily fluids takes place with an orderly motion. If the passage of the fluids is not interrupted by any blockage, then health persists; but an obstruction, brought about by a blockage of the corpuscles, causes diseases.

The lack of any mention of void in this passage is rather puzzling. Everything seems to be accounted for in terms of the elemental corpuscles—the bodily fluids, the impactions which interrupt their flow, and the pores themselves. The omission of void becomes still more striking when we reflect that void is also not mentioned in any non-doxographical accounts of the theory outside Galen. We could, of course, suppose that some notion of Epicurean-type void is implicit in the Caelius text which I discussed in Chapter 1, and in particular in the idea that the corpuscles are in constant motion. But the apparent ontological priority of corpuscles over pores (not to mention possible Empedoclean ancestry), which seems to me suggested quite strongly by Caelius, certainly militates against that bald atom/void, corpuscle/pore comparison which Galen promulgated so effectively.

No more is it necessarily likely that we are dealing with a physical theory like that of Strato. Lonie, following Diels, quite reasonably noted that the Asclepiadean pores could conceal a model of discontinuous void not unlike that of Strato; but Strato did not—or so it is generally thought—have fragile corpuscles in constant motion to contend with.[38] And another complication: the suggestion that the pores are naturally full of bodily fluids

---

[38] Lonie (1965), 128; Diels (1893).

going about their business means that they were not 'empty' but 'full'. Both Epicurus and Galen could happily have accommodated pores such as these in their physics; it seems very odd that Caelius is not more specific on such an important matter.

If Caelius and the other witnesses have nothing to say about void, then we are forced to return to Galen, who generally has a great deal to say. If we look more closely at the context of those passages where Asclepiades is being accused of selling out to Epicurus and negating continuum theory, some interesting results emerge. In each case the Methodists lurk not far away. In fact, Galen argues that the Methodists actually shared the Asclepiadean doctrine on void. In the *De causis morborum*[39] he sets his own view that diseases are 'simple and compound' against the doctrine of those people who hold that diseases result from 'an imbalance of the pores' and 'the dissolution of a perceptible unity'. These people, he says, hold that void 'is woven into every compound'. The doctrine looks like the mixture of Asclepiadeanism and Epicureanism we have become used to in Galen.[40] It emerges, however,[41] that 'these people' are Methodists; they are also made to believe that the body is compounded of ὄγκοι and πόροι.

Methodists like Soranus, of course, prided themselves on their lack of reliance on theory; the Methodists we know most about seem to have been just as scathing about the Asclepiadean theory as Galen was. It may well be the case that other

[39] 7. 1-2 K πόσα μέν ἐστι καὶ τίνα τὰ σύμπαντα νοσήματα κατ᾽ εἴδη τε καὶ γένη διαιρουμένοις, ἁπλᾶ τε καὶ σύνθετα, δι᾽ ἑτέρου δεδήλωται γράμματος. ἑξῆς δ᾽ ἂν εἴη τὰς αἰτίας αὐτῶν ἑκάστου διελθεῖν, ἀπὸ τῶν ἁπλῶν τε καὶ ὁμοιομερῶν ὀνομαζομένων τοῦ ζῴου μορίων ἀρξαμένους, εἶτ᾽ αὖθις ἐπὶ τὰ σύνθετά τε καὶ ὀργανικὰ μεταβάντας. ἐπεὶ τοίνυν ἐδείχθη, κατὰ μὲν τοὺς ἡνῶσθαί τε καὶ ἠλλοιῶσθαι τὴν ὑποβεβλημένην οὐσίαν γενέσει καὶ φθορᾷ δοξάζοντας, ἅπασα νόσος ὁμοιομεροῦς τε καὶ ἁπλοῦ πρὸς αἴσθησιν σώματος ἤτοι δυσκρασία τις ὑπάρχουσα, ἢ τῆς συνεχείας αὐτοῦ τῶν μερῶν διαίρεσις, κατὰ δὲ τοὺς μήθ᾽ ἡνῶσθαι καὶ κενόν τι παραπεπλέχθαι πάσῃ σώματος συγκρίσει νομίζοντας, ἀμετρία τε πόρων οὖσα καὶ λύσις τῆς αἰσθητῆς ἑνώσεως ἀρξώμεθα καὶ νῦν ἐπισκοπεῖσθαι, τὰς αἰτίας ἑκάστου τῶν νοσημάτων τῆς πρώτης ὑποθέσεως, ἣν δὴ καὶ ἀληθῆ πεπείσμεθα ὑπάρχειν.

[40] Although saying that void is 'woven into' every compound seems odd: Gottschalk draws a distinction between the verbs παραπλέκω, used here and diagnostic of Strato's interstitial void, and περιπλέκω, more associated with the Epicurean theory. παραπλέκω is certainly the *lectio difficilior* here.

[41] 7. 32-3 K.

Methodists like Thessalus were not the same, but it seems likely, once again, that Galen has more on his mind than he explicitly admits to. As it stands, the best and perhaps the only way to come to grips with the pores is to see how they functioned in Asclepiades' dynamics. And this brings us to the theory of 'movement towards what is fine'.

Theories set up to explain the movement of liquids and fluids generally in the body are legion in Greek medicine. That is no surprise; such theories had obvious applications in accounts of basic functions such as respiration, digestion, and conception. They were also used to explain the absorption of drugs in the body, and phenomena such as bruising and inflammation. Most of these models of movement were still in active use, or at least known, in Asclepiades' day.

The naturally attractive properties of certain internal structures were well known to Hippocratic doctors.[42] These properties were illustrated with a variety of analogies which the reader might have been able to test for himself (either in practice or in thought). Yet there was not, as a rule, an explicit, fully fledged theory behind these assumptions. The significance of particular shapes, for instance, comes up in the Hippocratic treatise *De morbis* 4; here the author appeals to the attractive power of empty space.[43] The use of this concept may well prefigure its appearance (as *horror vacui*) in the physics of Strato and Erasistratus. That, certainly, seems more likely than any hint of Democritean atomism.

Other theories of attraction implicated heat. Heat and space together are used by Aristotle to describe how the uterus draws in semen: the uterus is like a conical flask, which, when washed out with hot water and inverted in water, draws the liquid up.[44] Heat and space also figure in Hero's (and probably Strato's) explanation of how cupping-glasses work.[45] Cupping-glasses, a

[42] Cf. the Hippocratic *De vetere medicina* 22. (1. 626–34 L).

[43] *De morbis* 4. 35 (= 7. 548 L). Lonie (1981 a), 266, discusses its use in this treatise, and the similarities between this and *De natura pueri* 19. 1 (7. 506 L), 22. 4 (7. 516 L), 26. 3 (7. 526 L).

[44] *De generatione animalium* 739[b].

[45] See Gottschalk (1965), 109. 18–110. 2.

favourite ancient example of attraction at work, also figure in Plato's theory of $\pi\epsilon\rho\acute{\iota}\omega\sigma\iota\varsigma$, which accounts for the movement in terms of a 'pushing' of matter brought about by pressure exerted by the surrounding continuum.[46]

Asclepiades is credited by Galen with the idea that fluids tend to flow towards areas of 'fineness', or lower density, called variously $\lambda\epsilon\pi\tau o\mu\epsilon\rho\acute{\epsilon}\varsigma$ and $\lambda\epsilon\pi\tau o\mu\acute{\epsilon}\rho\epsilon\iota\alpha$. The notion had a wide application in the Asclepiadean accounts of respiration, digestion, pulsation, to name just a few. Galen is the only source to name this movement $\pi\rho\grave{o}\varsigma$ $\tau\grave{o}$ $\lambda\epsilon\pi\tau o\mu\epsilon\rho\grave{\epsilon}\varsigma$ $\phi o\rho\acute{a}$, but he evidently thought the phrase so strange that we may reasonably assume that it was used by Asclepiades. He also seems to have regarded the concept as a corner-stone of Asclepiades' pathology. If one were to write a commentary on a book by Asclepiades (as one might do with Hippocrates), he says, then the pores, the corpuscles, and $\pi\rho\grave{o}\varsigma$ $\tau\grave{o}$ $\lambda\epsilon\pi\tau o\mu\epsilon\rho\grave{\epsilon}\varsigma$ $\phi o\rho\acute{a}$ would need to be examined in detail.[47] Here, puzzlingly, no mention of void.

The terminology on its own gives little away. $\phi o\rho\acute{a}$ is a general term for physical movement. It *can*, needless to say, refer to the movement of the Epicurean atoms,[48] but can equally be opposed to the type of movement thought to be inherent in qualitative change ($\dot{a}\lambda\lambda o\acute{\iota}\omega\sigma\iota\varsigma$). In that sense, it can also be opposed to the idea of dynamically attractive movement, $\dot{o}\lambda\kappa\acute{\eta}$.[49] If $\phi o\rho\acute{a}$ tells us little, $\lambda\epsilon\pi\tau o\mu\epsilon\rho\acute{\epsilon}\varsigma$ is only slightly more helpful. That term too is essentially indifferent between atomism and continuum theory.[50] But in the context of a corpuscular theory we are justified in understanding the term etymologically, and in supposing that 'the fine' is made up of small particles. It is only a small step from there to imagining that $\lambda\epsilon\pi\tau o\mu\epsilon\rho\acute{\epsilon}\varsigma$ is descriptive simply of the size of the particles and not of their own

[46] *Timaeus* 79B and Gottschalk (1965), 132.

[47] *Commentarium III in Hippocratis librum III Epidemiarum* 17A. 506 K.

[48] *Ad Herodotum* 61.

[49] A concept without a place in Epicurean physics. See Galen, *De naturalibus facultatibus* 2. 2–4 K; *De locis affectis* 8. 32 K. The distinction was an old one; see Aristotle, *De generatione et corruptione* 319^{b}32–320^{a}7.

[50] Furley and Wilkie (1984), 260 n. 8, make the point well. Plato can deny void, and account for low density quite cheerfully (e.g. *Timaeus* 58A–B).

internal nature. It could, I suppose, be thought that corpuscles which form 'the fine' are less dense in some way than others, but there is no evidence that Asclepiades distinguished his particles in any such way. The Epicurean lobby would be pleased at this news; this language closely parallels the Epicurean discussion of the nature of the soul.[51] The example of the soul is the subject of that by now familiar passage in Calcidius where Asclepiades' corpuscles are also mentioned.[52]

Galen takes this view:

τὸν ἀέρα τοίνυν οὐκ ἔστιν ὅστις οὐκ εἶπε λεπτομερῆ, τῷ καταθραύ-
εσθαι δηλονότι ῥᾳδίως εἰς λεπτὰ μόρια καὶ διὰ πυκνοτάτων σωμάτων
ἑτοίμως διέρχεσθαι ἤ, εἴπερ ἐξ ἄλλου τινὸς ἐπιφέρουσιν αὐτῷ τὸ
λεπτομερὲς ὄνομα διδασκόντων ἡμᾶς σαφῶς. οὐ γὰρ δὴ ἐξ ὄγκων γε
λεπτῶν, ὡς οἱ τῆς ἑτέρας αἱρέσεως εἴποιεν ἂν ἡγεμόνες, ἐροῦμεν καὶ
ἡμεῖς συγκεῖσθαι τὸν ἀέρα. συνεχὴς γάρ ἐστιν καὶ εἷς ὅλος, οὐδαμόθι
κενὸν ἐν ἑαυτῷ περιέχων οὐδέν.

No one has ever denied that air is fine because it can readily be broken down into fine parts which easily penetrate the most solid bodies—if this is not the sense in which the term is applied, then let them make that clear to us. For we would not agree that air is composed of fine ὄγκοι (as the leaders of the other sect might say). We maintain that it is continuous, a single whole, and that it nowhere contains any void at all.[53]

He evidently takes it for granted that the presence of λεπτο-μέρεια need not presuppose the existence of void, unless it is coupled with the idea of λεπτομέρεια. Caelius glosses Asclepiadean λεπτομερές as *spiritus* (i.e. 'pneuma'), and in several places suggests that it is to be connected with *fervor*; the overall impression which we gain from Caelius is that λεπτομερές refers simply to very fine corpuscles.[54] Elsewhere, Cassius the Iatrosophist notes that in cholera pneuma is carried towards the

---

[51] e.g. *Ad Herodotum* 63.

[52] *In Timaeum*, pp. 229–30 W.

[53] *De simplicium medicamentorum temperamentis ac facultatibus* 11. 423 K.

[54] e.g. at *De morbis acutis* 1. 119 'ad cerebri membranam spiritum vel fervorem ferri' is evidently a description of PTLP at work; at *De morbis chronicis* 3. 65 the *tenuissima corpuscula* are glossed by Caelius as 'spiritus, quem λεπτομέρειαν eorum princeps [sc. Asclepiades] appellavit'.

λεπτομέρεια in the belly because the belly is warm.[55] The implication is that some corpuscles are larger and more dense than
others. Yet the 'leaders of the other sect' in Galen's words are
Methodists too; the view they are made to espouse looks clearly
Asclepiadean. All this does not help very much at first. Many
people besides Asclepiades (and including continuum theorists
as well as Epicureans) who held that the soul was made up of
pneuma might be expected to describe it as λεπτομερές.[56]
Indeed, whether you were a continuum theorist or not you could
expect to find yourself using a certain amount of quasi-technical
terminology like this in common with people of quite different
persuasions.[57]

For all these people, λεπτομέρεια carried with it connotations
of penetrative power and mobility. Galen often explains the
powers of drugs which can easily permeate the whole body in
terms of their 'fineness', and the terminology of λεπτομέρεια is
extremely common in Galen's pharmacological works.[58] Against
this kind of background we might imagine that Asclepiades used
the term to describe 'rarity' or permeability generally.

So this terminology too, like that of the pores, does not greatly
assist in our search for void. Yet it is not entirely useless. Galen's
attack on Asclepiades in the *De naturalibus facultatibus* is the
most important source for our knowledge of PTLP. The argument

---

[55] *Problemata* 72 διὰ τί ἐν ταῖς χολέραις τὰ ἄκρα συνέλκεται καὶ σπᾶται καὶ κατα-
ψύχεται, καὶ ἀμαυρὸν τὸν σφυγμὸν ἔχουσιν; ὅτι λεληθότως διαφορεῖται τὸ ἐξ αὐτῶν
πνεῦμα πρὸς τὴν ἐν κοιλίᾳ λεπτομέρειαν. ἔνθερμος γὰρ τούτοις ἡ κοιλία.

[56] For just a few examples see Diocles *ap.* Vindicianus, *De semine* 41 = Wellmann
(1901), 233; Democritus, Epicurus, and Asclepiades *ap.* Calcidius, *In Timaeum*, pp. 229–30
Waszink. 'The Stoics' (*sic*) ps.-Galen, *Definitiones medicae* 19. 355 K. For the Asclepiadean
view that animals breathe in order to generate and sustain the soul, see Galen, *De usu
respirationis* 4. 471, 483–4 K; *Commentarium VI in Hippocratis librum VI Epidemiarum* 17 B.
246 K; ps.-Galen, *De historia philosophica* 24.

[57] Contrast the very different applications of λεπτομερές at Galen, *De simplicium
medicamentorum temperamentis ac facultatibus* 11. 516–17 K, and *An in arteriis* 4. 705 K, with
that at Hero of Alexandria, *Pneumatica*, pp. 26–8 Schmidt—all cases where it is used of
'lightness', though against very different theoretical backgrounds. Both a continuum
theorist and an atomist could say that something weighs little on account of the λεπτομέρεια of its parts. This usage goes back a long way; cf. Theophrastus, *De sensu* 62, on
Democritus.

[58] e.g. *De compositione medicamentorum secundum locos* 12. 387, 399, 441, 667 K, and *Ars
medica* 1. 384, 388 K.

there centres on Galen's insistence that Asclepiades discarded traditional medicine altogether by 'denying that anything is attracted by anything else'. He spends some time gleefully underlining what he sees as a major disagreement between Epicurus and Asclepiades over how 'magnetic' attraction should be explained. Epicurus, claims Galen, gave an explanation of this phenomenon in terms of the interweaving of atoms which flow out from the magnet and from the object attracted. But Asclepiades, even though he posited elements similar to those of Epicurus, rejected this explanation, without even bothering to substitute another. In doing so, Asclepiades effectively denied the existence of attraction in nature, let alone in the body, or so Galen concludes.[59]

Galen's conclusion may sound bizarre, even if by now we are becoming used to reading between the lines of what he writes. To understand Galen at all, we need to recognize his own fundamental doctrine of attraction and its different species. In fact, Galen shows his true physiological colours when he distinguishes between different types of attraction at *De naturalibus facultatibus* 2. 206–7 K. The first type he calls πρὸς τὸ κενούμενον ἀκολουθία (PTKA). This type corresponds to what we might call *horror vacui*, and it is particularly associated in our sources with Erasistratus. In some of its guises, this type of explanation was accepted by Galen. The second type of attraction, more important for Galen's own physiology, is called 'assimilation to what is appropriate'. The phenomenon of magnetism is Galen's paradigm for this type, whilst the way in which air is drawn into bellows illustrates PTKA. 'Attraction to what is appropriate' is an impossibly vague heading in practice; in fact Galen includes in this category types of theory which might be thought to belong better in the first class. Take, for example, the attractive power of the 'openness' of the mouth, used to explain appetite.[60] It is the 'suitability' of the shape of

---

[59] *De naturalibus facultatibus* 2. 44–6 K. Significantly perhaps, Strato seems to have levelled much the same charge at Plato: see Simplicius, *In Physicorum*, p. 663. 6 Πλάτων αὐτὸς τὴν ἑλκτικὴν δύναμιν ἀναιρεῖν δοκεῖ.

[60] εὐρυχωρία at *De naturalibus facultatibus* 2. 174 K.

the mouth which induces attraction of food, rather than any hint of *horror vacui* at work.[61] If 'magnetism' is for Galen an example of his second, favoured type of attraction then we can see a little more clearly why he is so concerned that Asclepiades does not share the Epicurean explanation of magnetism. Epicurus at least provided an explanation falling very broadly into Galen's second category. Asclepiades did not. This is a recurring theme in the *De naturalibus facultatibus*; Asclepiades is portrayed as a pedantic figure far more concerned with 'logical consistency' with his own principles than with accounting for the phenomena.[62]

So far I have been taking Galen at his word. There is no guarantee that Asclepiades did in fact disagree explicitly with Epicurus' account of attraction, nor, indeed, that he would have spoken about attraction in the way Galen suggests that he did. We must pursue more indirect, discreet lines of inquiry. Most promising among these, I believe, is that suggested by the similarity between PTKA and PTLP. Since the examination of this similarity will take up much of what follows, I should introduce a few cautionary remarks. In one sense, the expression PTKA is not explicitly attested for Erasistratus, any more than ἄναρμοι ὄγκοι is for Asclepiades. In fact, one modern scholar has doubted that the attribution is necessarily correct.[63] But it is found so widely in Erasistratean contexts in Galen, where it assumes a status parallel to other Erasistratean technical terms, that I feel we are justified in assuming its authenticity.[64]

PTKA and PTLP had a similar range of applications. Both Erasistratus and Asclepiades used their respective theories to explain the pulse,[65] appetite, digestion and the assimilation of

---

[61] This is an important objection to those (like Wellmann (1929)) who see in the Hippocratic use of terms such as εὐρυχωρία the signs of a void theory.

[62] e.g. *De naturalibus facultatibus* 2. 50–1 K.

[63] Von Staden (1975).

[64] See below, n. 67; note also Galen's passing remark in his commentary on Plato's *Timaeus* 3. 17: γίνεσθαι δέ φησιν ὁ Πλάτων τὴν περίωσιν ταύτην διὰ τὴν ἀνάγκην τοῦ κενοῦ, τούτεστι τοῦ μηδεμίαν χώραν γενέσθαι κενήν· καὶ διὰ τοῦτο τοῦ κενουμένου τὸ συνεχὲς ἔπεται τὴν χώραν αὐτοῦ πληροῦν, ὅπερ Ἐρασίστρατος ὀνομάζειν εἴωθε τὴν πρὸς τὸ κενούμενον ἀκολουθίαν.

[65] Erasistratus: Galen, *In Timaeum* 3. 19; Asclepiades: *De differentiis pulsuum* 8. 748 K.

nutriment,[66] the secretion of bile[67] and urine.[68] Both theories, as we shall see in the next chapter, were central to their authors' pathology; PTLP explains the movement of corpuscles to places where they cause diseases, while PTKA informed the similarly fundamental transfusion of blood into the arteries which made Erasistratean patients sick.[69] The primary sources for PTKA[70] have been interpreted in a variety of ways, and this calls for a brief bibliographical interlude. The classic account is that of Diels (1893); his view is that PTKA looks back to Strato's concept of *horror vacui*. Doubts about this were expressed, as I said at the beginning of Chapter 1, first by Schmekel (1938), who suggested that the main source for our knowledge of Strato, Hero of Alexandria, is more problematic than was first thought. His doubts were developed by Gatzemeier (1970). More recently still, Furley[71] has underlined the difficulties that have arisen through the failure of some modern scholars to distinguish between different types of *horror vacui*. An important school of thought in Germany, particularly associated with Wellmann, maintained that Erasistratean PTKA had nothing whatever to do with Strato.[72] The Wellmann line connected PTKA ultimately with Plato's 'circular thrust' or περίωσις, via Erasistratus' teacher Chrysippus. This view has received little support in recent years, but still deserves consideration.

[66] Appetite and digestion (Erasistratus): *De naturalibus facultatibus* 2. 104–5 K; (Asclepiades): Caelius, *De morbis acutis* 1. 113–14.

[67] Erasistratus: *De naturalibus facultatibus* 2. 63 K; Asclepiades: ibid. 2. 39–40 K.

[68] Erasistratus (?): *De naturalibus facultatibus* 2. 77–8 K; Asclepiades: ibid. 2. 31–2 K. These last two cases are slightly more doubtful. At ibid. 2. 77–8 K Galen implies that Erasistratus used PTKA to explain urinary secretion, but at 2. 63–4 K he suggests the opposite. At ibid. 2. 187 K he says that both Erasistratus and Asclepiades committed the same errors in this respect, that is to say in depriving the stomach, womb, and bladder of their 'natural facilities'.

[69] See ch. 4.

[70] The most important include *De naturalibus facultatibus* 2. 63, 76–7, 95–106, 204 K; *De usu partium* 3. 492 K; *An in arteriis* 4. 710–14 K; *Anatomicae administrationes* 2. 649 K; *De venae sectione adversus Erasistratum* 11. 154, 324 K; Anonymus Londinensis 26. 31–28. 45. Most of these passages are discussed by Fuchs (1892*a*); see also the Erasistratean aetiology of paralysis at Fuchs (1894*a*), 550. Among passages not mentioned by Fuchs, Palladius, *Commentarium VI in Hippocratis librum VI Epidemiarum* 2. 151 Dietz draws attention to the anti-teleological consequences (as far as Galen is concerned) of such a theory.

[71] Furley and Wilkie (1984); and Furley (1989).

[72] Wellmann (1900), 371, and in PW s.v. 'Erasistratos'.

The origins of PTKA, let alone PTLP, are more complex, I suspect, than any of these models might suggest. For a start, we must not underestimate Erasistratus' own intellectual independence.[73] On one hand we have the well-attested links between Erasistratus and the Peripatos. And here the testimony is not just Galen's. Diogenes Laertius reports that Erasistratus studied with Theophrastus: this implies at the very least that he was acquainted with scientific developments in the Lyceum.[74] It is not difficult to understand the popularity of Diels's thesis that there is a link between PTKA and the Lyceum. But there is some very powerful evidence to suggest that Erasistratus was closely involved in scientific issues raised by members of the Academy. Hardly surprising perhaps; after all, the *Timaeus* was one of the most influential scientific texts in later antiquity. Galen, in his commentary on the *Timaeus*, refers more than once to Erasistratus' elucidation of the notorious ἐγκύρτια which figure in the account of respiration at *Timaeus* 79A. A number of later scientists in the Academy tried valiantly to discover what Timaeus was talking about.[75] At 3. 17–18 of Galen's commentary, Hestiaeus' modifications to the explanation are considered,[76] and Galen seems explicitly to compare PTKA with περίωσις. Later, Galen tells us that Erasistratus actually refuted the argument that respiration can be explained in terms of the Platonic model. I shall deal with the specific issues raised here in more detail when I come to consider Asclepiades' theory of

[73] Galen claims that Erasistratus followed his teacher Chrysippus 'in all things' at *De venae sectione adversus Erasistratum* 11. 197 K (cf. 171 K), but at *De placitis Hippocratis et Platonis* 5. 185 K he points to a fundamental divergence between teacher and pupil over the origin of psychic and inspired pneuma. Galen loves to present slavish adherence to the precepts of one person as a typical defect in his enemies.

[74] Diogenes Laertius 5. 57; compare the rather confused notice of Pliny at *Historia naturalis* 29. 5 'ex Chrysippo discipulus eius Erasistratus Aristotelis filia genitus'.

[75] The explanation of respiration in the *Timaeus* is opaque. Plato's analogy involves some kind of revolving woven object in the shape of a wheel which has funnels (ἐγκύρτια) around it. The problem here is whether the wheel makes half or full turns and how this corresponds to the respiratory cycle.

[76] Hestiaeus' position, as Galen sees it, seems to be that inspired air enters the body through the skin, passes out through the mouth and nose, then re-enters through the mouth and leaves through the skin. Modern scholars tend to reject this interpretation. See Fritzsche (1902), 376; Taylor (1928), 562–9; Lloyd (1966), 360.

respiration; what is important at this point is the simple observation that Erasistratus was involved in the debate in the first place. This does not, of course, necessitate a close involvement with the Academy. In fact it probably supports those notices which link him with the Lyceum; Theophrastus' *Metaphysica* shows a growing awareness of the need to explain, and where necessary resist, the scientific doctrines of the Academy after Plato.[77] Erasistratus fits well in this tradition, and so, I shall suggest, does Asclepiades.

As Galen would have it, PTKA simply suggests that when a hollow body is evacuated, something must enter it to take the place of what has been removed. What enters must be equal in amount to what was evacuated.[78] One specific application of the principle is in the explanation of the absorption of nutriment into the veins, ἀνάδοσις. In line with Erasistratean PTKA, the veins are kept filled whenever anything flows out from them because 'there are only two things which could happen'. The loss of material from these vessels could either result in the formation of an unnatural 'large-scale' vacuum, or in the attraction of the matter which is contiguous to the veins.[79] The first case, says Galen, is an impossibility for Erasistratus, since he insisted that a vacuum cannot occur in nature. It is not altogether clear at this stage whether or not Erasistratus would have accepted the possibility of a vacuum brought about by force, though it should be noted that Galen attributes to Erasistratus the language which we associate especially with the testimonia relating to Strato. Galen is disinclined to clarify the point, and says nothing of substance about Erasistratus' stance on the existence of disseminate void. *Horror vacui* (and PTKA as I have described it, for that matter) could be acceptable to a strict continuum theorist. To such a person the mere threat of a large-scale unnatural void would be enough to ensure that it does not happen. To one who accepts the existence of disseminate void at the level of elemental particles, there need be no difficulty either

---

[77] As I have argued in Vallance (1988).
[78] Galen, *De usu respirationis* 4. 473-4 K; and cf. *De locis affectis* 8. 324-5 K.
[79] I am following the account at *De naturalibus facultatibus* 2. 75-6 K.

as long as some kind of distinction is made between theoretical, interstitial void (which exists simply because the component parts of matter do not fit together perfectly) and large-scale void.[80]

As far as Galen is concerned, PTKA cannot work where disseminate void exists. In the *De usu respirationis* and the *De naturalibus facultatibus* (in passages which I shall consider shortly), he seems to have picked up an argument that where disseminate void is present, unnatural vacuum cannot occur, since the underlying elements simply spread out and rarify when matter is evacuated from an enclosed space. If the particles in such a physical model are really interrupted by void spaces, an active vacuum or even the threat of one is not possible. Granted that Galen accepts the validity of PTKA, in some cases at least, it may seem surprising that he is unaware of the Stratonic arguments for *horror vacui* coexisting with disseminate void, especially given Erasistratus' supposed reliance on this tradition.

I wish to consider two 'test cases' of PTKA at work. The first application of the principle is outlined at *De usu respirationis* 4. 473-4 K. The text and translation belong to Furley and Wilkie.

ἀλλὰ ... ὁ Ἐρασίστρατός φησιν, ὅτι μηδ' ἕλκειν δύναται τὸν ἐκ τοῦ πνεύμονος ἀέρα κατὰ τὴν τῆς ἀναπνοῆς ἐπίσχεσιν ἡ καρδία· φυλάττεσθαι γὰρ τῶν ἀναπνευστικῶν ὀργάνων τὸν ἴσον ὄγκον τῆς διαστάσεως ἐν ταῖς τοιαύταις καταστάσεσιν. εἴπερ οὖν εἵλκυσέ τι μέρος ἀέρος ἡ καρδία, κενὸς ἂν ὁ τοῦ μεταληφθέντος ἐγένετο τόπος· τοῦτο δὲ γένεσθαι ἀδύνατον. ἵν' οὖν μὴ γένηταί τι ἀδύνατον, οὐδὲ μεταλήψεσθαί φησι τὴν ἀρχήν. δεῖ γάρ, ἵνα τι μεταληφθῇ, μὴ μόνον εἶναι τὸ ἕλξον, ἀλλὰ καὶ τὸ μεταδῶσον· οὐ μεταδίδωσι δ' ὁ θώραξ, τὸν ἴσον ὄγκον φυλάττων τῆς διαστάσεως· οὐδ' οὖν οὐδ' ἡ καρδία δύναται μεταλαμβάνειν, ἀλλ' ἐπιχειρεῖ μὲν ὡς ἔμπροσθεν, ἀνύει δ' οὐδέν· κἀντεῦθεν τὸ πνίγεσθαι.

---

[80] This distinction is brought out in Hero of Alexandria's example of how one can create an artificial vacuum by force, sucking the air out of a pot: *Pneumatica*, p. 8 Schmidt. The presence of disseminate void in no way interferes with this explanation, as far as Hero can see. If some modern scholars have been unwilling to accept that *horror vacui* is not in itself diagnostic of either atomism or continuum theory, the ancients could be similarly confused. See ps.-Alexander's explanation of what happens when we suck wine through a tube, at *Problemata* 2. 59.

But . . . Erasistratus says that the heart cannot even draw the air out of the lung during the stoppage of breathing, because in these conditions the same volume of expansion is maintained by the organs of breathing. Hence, if the heart had taken some part of the air, the place of the air that was taken would have become empty; and that is impossible. So that, therefore, the impossible may not happen, he says that it [the heart] will not even take any of it in the first place. For that something may be taken, it is not enough that there should be something that will attract, but there must also be something that will give; and the thorax does not give, retaining as it does the same volume of expansion; neither, therefore, can the heart take; but it tries as it did before, and gets nothing, and hence the suffocation.

Respiration, for Erasistratus, is a matter of the thorax expanding and drawing in air ultimately to the lungs by means of ΡΤΚΑ.[81] The thorax itself derives its own movement from the muscles (477 K). From the lungs, inspired air is drawn into the heart, and from there it is pumped into the arteries. Here Erasistratus is faced with a problem: when we hold our breath, the heart still 'beats' (i.e. it is still pumping something), even though it can no longer be drawing in air from the lungs because this would involve the formation of an unnatural void there. It is this impasse which, according to Erasistratus, causes suffocation.

Galen evidently feels that this also suffocates Erasistratus' theory. His reasons, as ever, are not what they might seem. What follows on from this passage is a set of two distinct and mutually exclusive refutations of Erasistratus' explanation of suffocation. The first argument *contra Erasistratum* is attributed to an unidentified group of 'refutationists'. Again the text and translation are from Furley and Wilkie.

ταυτὶ μὲν ὁ Ἐρασίστρατος. οἱ δ' ἀντιλέγοντες αὐτῷ πρῶτον μέν, ὅτι
οὐδ' ἀπέδειξέ που μηδ' ὅλως παρεσπάρθαι τοῖς σώμασί πού κενόν,
ἀλλ' ὅτι μὴ ἀθρόον, ὑπομιμνήσκουσιν, ἔπειτα δ', εἰ καὶ τοῦτο
συγχωρηθείη, τὸ μηδ' ὅλως ἐν τῷ κόσμῳ μηδαμοῦ παραπεπλέχθαι
κενόν, ἀλλ' οὔ τι γοῦν ἀδύνατον εἶναί φασι τὴν αὐτὴν οὐσίαν χεομένην

---

[81] For further details see Anonymus Londinensis 23; *De placitis Hippocratis et Platonis* 5. 185 K, on the type of inspiration involved; *De locis affectis* 8. 429–30 K, on the activities of the muscles. The best modern account is that of Furley and Wilkie (1984), 26–37.

τε καὶ πάντῃ τεινομένην μείζονα τὸν πρόσθεν τόπον ἐπιλαμβάνειν, καὶ
τρίτον, ὅτι τοῖς ἐναργέσιν ὁ λόγος αὐτοῦ μάχεται· λέγει μὲν γάρ, ὡς
οὐχ ἕλκει κατὰ τὰς ἐπισχέσεις τῆς ἀναπνοῆς ἡ καρδία τὸν ἐκ τοῦ
πνεύμονος ἀέρα· τούτῳ δ' ἕπεται τὸ μὴ διαστέλλεσθαι τὴν καρδίαν
(οὐδὲ γὰρ ἐνδέχεται κατ' αὐτὸν διαστέλλεσθαι μέν, μηδὲν δ' ἕλκειν·
κενὸς γὰρ ἂν οὕτω γένοιτο τόπος)· εἰ δ' οὐ διαστέλλεται, δῆλον ὡς
οὐδὲ κινεῖται· ἀλλὰ μὴν φαίνεται κινουμένη· τὸ ἐναντίον ἄρα τοῦ
πρώτου περαίνεται, τὸ τὴν καρδίαν ἕλκειν ἐν ταῖς ἐπισχέσεσι τῆς
ἀναπνοῆς τὸν ἀέρα.

So far Erasistratus. But his opponents recall that in the first place, he
has nowhere proved that void is not at all interspersed anywhere in
solid bodies, but only that it is not present *en masse*; and second, that
even if this be conceded, namely, that void is nowhere at all inter-
woven in the universe, even so, they say, it is not impossible for the
same substance, relaxing and stretching in all directions, to take up a
greater space than formerly; and third, that his opinion is in conflict
with what is evident to the senses. For he says that the heart does not
draw air from the lungs when the breath is held; and it follows from
this that the heart is not expanded (for it is not possible according to
him that it should expand, yet draw in nothing; for an empty space
would thus arise); and if it does not expand, clearly it does not move.
But it evidently does move; so the conclusion is the contradictory of
the original thesis, namely, that the heart does draw in air when the
breath is held.[82]

The opponents of Erasistratus are making two main points
which concern me here. Firstly, in saying that the heart can
draw nothing from the lungs when the breath is held, Erasistratus has not considered the possibility of the lungs' contents
expanding or stretching. This, they say, is a crucial point regardless of whether or not the existence of void is accepted. PTKA
may involve the denial of extended or large-scale void, but
Erasistratus has nowhere accounted for the possibility of void
interstices in matter which would, in the view of his opponents,
allow for rarefaction. Secondly, it is implied, rightly or wrongly,
that Erasistratus denied the existence of *any* type of void. But in
doing this, he has also been inconsistent as a continuum theorist,

[82] 4. 474-5 K.

since he ignored the arguments of continuum theory about the stretching and compression of matter.

The idea that positing disseminate void is damaging to PTKA (through its damage to the Erasistratean explanation of suffocation) seems strange at first sight; it was evidently not damaging to Stratonic *horror vacui*, where such void could only exist in interstices. The arguments of these opponents of Erasistratus could well be pointing to a rather different conception of what is meant by disseminate void. But we must take seriously their claim that Erasistratus did not even discuss this kind of void.

Galen now proceeds to introduce some Erasistrateans who defend their master. They insist that the points about disseminate void and stretching of a continuum do not apply, and argue that when the breath is held, the heart's movement continues as normal, except that it gets its supply of air from the 'great artery'. At this point, another objection to Erasistratus is brought on; Galen does not signal its arrival very clearly:

τινὲς δὲ οὐδὲ διαστέλλεσθαι καὶ συστέλλεσθαι λέγουσιν αὐτήν [i.e. the heart], ἀλλ᾽ οἷον κραδαίνεσθαι. τὰ μὲν γὰρ ὑπὸ τῶν περὶ τὸν Ἀσκληπιάδην εἰρημένα κάλλιον εἶναί μοι δοκεῖ παραλιπεῖν, ἄτοπά τε φανερῶς ὄντα καὶ τῶν προσηκόντων ἐλέγχων ὑπ᾽ Ἀθηναίου τετυχηκότα.

But some say that the heart does not actually expand or contract, but only oscillates, as it were. As for the things said by the school of Asclepiades, I think them better passed over in silence, being clearly foolish, and having received the appropriate refutation from (?) Athenaeus.[83]

Furley's translation of the beginning of the second sentence above does not really make it clear that the first sentence refers to the Asclepiadean doctrine.[84] If this is correct, what is the force of this particular objection to Erasistratus' theory? The obvious

---

[83]   4. 475 K (trans. Furley and Wilkie).

[84]   In fact the passage is a difficult one textually; for *Athenaeus* the MSS read *Asclepiades*, but all modern editors have condemned this on the ground that Asclepiades should not be refuting his own doctrines. But there need be no problem; Galen could well be saying that the Asclepiadean doctrine is incoherent with Asclepiades' own principles.

effect of denying expansion and contraction to the heart will be to remove the heart's mechanical role as a pump. (A more accurate rendering of κραδαίνω might be 'quiver' or 'shake'.) It would appear that the Asclepiadeans, with whom Galen is here characteristically unimpressed, also have an argument which denies the possibility of attraction to the heart through PTKA.[85]

The second set of arguments against Erasistratus' theory of suffocation belongs to Galen himself, and it encompasses a refutation of what was said by the renegade Erasistrateans as well as the Asclepiadeans. Galen's arguments are somewhat diffuse and verbose, so I have summarized them as follows. Erasistratus and all the others were wrong because:

1. The heart cannot draw air from the 'great artery' (as the followers of Erasistratus maintain) because there are valves whose purpose is to inhibit this kind of reflux, and Erasistratus knew about them perfectly well.

2. The 'shaking of the heart' view is wrong because it is in conflict with the phenomena of cardiac expansion and contraction.

3. Both the above views are wrong, because the pulsation in the arteries is unchanged regardless of whether or not we hold our breath.

Granted Galen's constant juxtaposition of Erasistratus and Erasistrateans, Asclepiades and Asclepiadeans, it is not at all easy to distinguish what was actually believed by any one person.[86] Neither this passage, nor any of those in the *De naturalibus facultatibus* gives any clear idea of whether or not PTKA was

---

[85] Galen uses κραδαίνεσθαι again in the *De differentiis pulsuum* in a passage which I shall discuss presently; with the type of movement involved in 'shaking' or 'oscillation' cf. the way in which σείομαι is used of the lungs' motion at *De usu partium* 3. 466–7 K. (I wonder if an etymological connection between κραδαίνω and κραδίη (=heart) was perceived in antiquity.) Wellmann's attempt (1908: 687) to link this with the ultimately Democritean concept of φλεβοπαλία is unconvincing.

[86] Galen pretends that it was not easy for him. At *De venae sectione adversus Erasistratum* 11. 221 K he claims that in his own day all of the works of Erasistratus had perished, and that he had to rely on the evidence of Erasistrateans. This is surely tendentious. Aulus Gellius was able to quote verbatim from Erasistratus' *On Definitions* at *Noctes Atticae* 16. 3. 5–7. The Erasistrateans of Galen's own time included several successful men in Rome who were *medici non grati*. Smith (1979) gives a good overall view of this aspect of Galen's character.

part of a strict continuum theory, or one which admitted disseminate void. But in the light of the arguments in the *De usu respirationis*, the first might be more likely.

Aulus Gellius raises another problem. At *Noctes Atticae* 16. 3 he recalls the time when his friend Favorinus gave him the Erasistratean explanation of why it is that when one does not eat anything, hunger comes and then passes away after a few days:

nam quod Erasistratus scriptum, inquit [sc. Favorinus], reliquit propemodum verum est: esuritionem faciunt inanes patentesque intestinorum fibrae et cava intus ventris ac stomachi vacua et hiantia; quae ubi aut cibi complentur aut inanitate diutina contrahuntur et conivent, tunc loco, in quem cibus capitur, vel stipato vel adducto, voluntas capiendi eius desiderandique restinguitur.

Erasistratus was quite close to the mark, he said, when he wrote that the empty and gaping fibres of the intestines, the hollowness within the belly, and the vacuous and yawning parts of the stomach bring about hunger. But when they are either filled with food or contracted and brought together by continued abstinence, then the desire to take food and yearn for it is extinguished because the place which accepts food is either filled or contracted.[87]

A little further on, the problem and its explanation continues in Greek: why do the Scythians bind their bellies with tight belts to stave off hunger?

εἰθισμένοι δέ εἰσιν καὶ οἱ Σκύθαι, ὅταν διά τινα καιρὸν ἀναγκάζωνται
ἀσιτεῖν, ζώναις πλατείαις τὴν κοιλίαν διασφίγγειν, ὡς τῆς πείνης
αὐτοὺς ἧττον ἐνοχλούσης· σχεδὸν δὲ καὶ ὅταν πλήρης κοιλία ᾖ, διὰ τὸ
κένωμα ἐν αὐτῇ μηδὲν εἶναι, διὰ τοῦτο οὐ πεινῶσιν, ὅταν δὲ σφόδρα
συμπεπτωκυῖα ᾖ, κένωμα οὐκ ἔχει.

The Scythians are accustomed, whenever they need to go without food, to bind their bellies with broad belts so that the hunger will not upset them so much. It is pretty much the case both that whenever the belly is full *and* whenever it is strongly compressed hunger is absent because the belly has no vacuity in it.[88]

Hunger for Erasistratus, according to this account, is related to the κενώματα, or empty spaces, in the belly which can be

---

[87] 16. 3. 3.   [88] Ibid. 8.

filled with food or drink. Are these spaces void? (LSJ translates κένωμα as 'vacuum'). While the phrase 'empty and gaping fibres of the intestines . . .' might look suggestive of void, the terms could just as easily refer to 'empty' (in the continuous sense) passages through which the distribution of nutriment takes place. Of more interest, perhaps, is the explanation's resemblance to the Asclepiadean one, something I shall consider shortly.

By now it will be clear that there are serious κενώματα and inconsistencies in our evidence for PTKA. We cannot even be sure yet whether or not Erasistratus accepted the existence of disseminate void, far less use our knowledge of Erasistratus to illuminate Asclepiades. The assumption that Erasistratus necessarily belongs to the same tradition as Strato of Lampsacus certainly needs further examination. But one thing seems secure. PTKA was a theory of *horror vacui*, and Wellmann was wrong to disagree with *that*.

One possible reason for all this confusion could lie in antiquity. It seems likely that PTKA came in for some very close scrutiny, even from followers of Erasistratus not long after his death. If we return to the case of ἀνάδοσις, as Galen describes it in the *De naturalibus facultatibus*, we find that Asclepiades himself is presented as one of these dissenters. Asclepiades argues that Erasistratus' use of the principle to explain how nutriment enters the veins fails because there are not *two* alternatives as to what will happen when the veins are emptied, as Erasistratus had maintained, but three. (It will be remembered that Erasistratus had thought that nutriment is drawn into the veins to prevent the impossible formation of a large-scale vacuum in them.) PTKA might work, according to Asclepiades, if the coats of the veins are rigid. But they are not, so we should reasonably expect that the coats of the vessels will collapse.

This is a powerful argument against this particular application of PTKA, especially when we consider that most of the vessels in the body are not rigid.[89] Galen thought the argument cogent

[89] Asclepiades believed that the arteries are naturally in systole. Contrast the Erasistratean discussion of morbid transfusion at Anonymus Londinensis 25. 25-8. 45. The

enough to adopt (or adapt) it for himself in the *An in arteriis* without mentioning Asclepiades at all. A rare case indeed.[90] Galen removes the distinction made in the *De naturalibus facultatibus*, between rigid vessels, where PTKA *will* work, and soft ones; he argues that the collapse should take place even with rigid vessels. All these observations lead us to two very important general conclusions. First, we may be fairly sure that Asclepiades was involved in some kind of debate about PTKA. Second, it is very unlikely indeed that Asclepiades' own PTLP was identical to PTKA. Clearly, this is not enough on its own to support my argument. Even if it is granted that Asclepiades' criticism of PTKA at *De naturalibus facultatibus* 2. 75-6 K is a cogent one, we are a long way from being in a position to prove that PTLP did in fact grow out of PTKA.

At *De naturalibus facultatibus* 2. 97-102 K Galen discusses another dispute, this time between two groups of warring Erasistrateans. It revolved around the ultimate constitution of their master's elemental τριπλοκία τῶν ἀγγείων, or 'threefold web' of vein, artery, and nerve which made up the Erasistratean body. Once again, none of these Erasistrateans is named by Galen; as far as he is concerned, they are merely convenient vehicles for his demonstration that even Erasistrateans could not agree about what Erasistratus said.

The first group, favoured by Galen, claim that the 'elemental nerve' of Erasistratus is absolutely continuous—otherwise, they say, why would he have said it was elemental?

ἠβουλόμην δ' αὖ πάλιν μοι κἀνταῦθα τὸν Ἐρασίστρατον αὐτὸν ἀπο-
κρίνασθαι περὶ τοῦ στοιχειώδους ἐκείνου νεύρου τοῦ σμικροῦ, πότερον
ἕν τι καὶ συνεχὲς ἀκριβῶς ἐστιν ἢ ἐκ πολλῶν καὶ σμικρῶν σωμάτων,
ὧν Ἐπίκουρος καὶ Λεύκιππος καὶ Δημόκριτος ὑπέθεντο, σύγκειται.
καὶ γὰρ καὶ περὶ τούτου τοὺς Ἐρασιστρατείους ὁρῶ διαφερομένους. οἱ
μὲν γὰρ ἕν τι καὶ συνεχὲς αὐτὸ νομίζουσιν ἢ οὐκ ἂν ἁπλοῦν εἰρῆσθαι

Anonymus argues that the vessels in the body are ἀσύμπτωτοι (27. 17-21), but adds that if we accept that they *are* collapsible, then under PTKA they should be expected to collapse. If this is an Asclepiadean view, Asclepiades himself is not mentioned.

[90] 4. 710 K εἰ δὲ μέχρι ποσοῦ τινος, ἐκεῖνος αὖθις ὁ λόγος ἀποδειχθήσεται μοχθηρός, ἐὰν τὸ κατὰ τοὺς ἁπαλοὺς καλάμους ἀναμνησθῶμεν φαινόμενον.

πρὸς αὐτοῦ φασι· τινὲς δὲ καὶ τοῦτο διαλύειν εἰς ἕτερα στοιχειώδη τολμῶσιν.

At this point, again, I should like Erasistratus himself to answer regarding this small elementary nerve, whether it is actually one and definitely continuous, or whether it consists of many small bodies, such as those assumed by Epicurus, Leucippus, and Democritus. For I see that the Erasistrateans are at variance on this subject. Some of them consider it one and continuous, for otherwise, as they say, he would not have called it simple; and some venture to resolve it into yet other elementary bodies.[91]

Galen can be seen here at his most disingenuous; he clearly believed that Erasistratus himself would have said it was continuous, since he says as much explicitly in a somewhat less highly charged context at *De naturalibus facultatibus* 2. 211 K.[92] He leaves the question open in the present passage, and brings in the opposing group, who argue that the 'elemental nerve' should be further dissolved into elementary bodies.[93] Galen is at pains to distinguish the Erasistrateans from Erasistratus here; his reasons are largely, if not wholly, polemical. Neither of their positions wins Galen's support. He argues (not altogether convincingly) that if we accept the first position, then the Erasistratean theory of how the 'nerve' itself is nourished falls apart.[94] This explanation apparently relied on PTKA just as did the theory of ἀνάδοσις on a larger scale. But the second position is particularly displeasing for Galen. If we accept it, he says, then we arrive at the elements of Asclepiades, and these contain no role for PTKA at all. This, I believe, is the source of

---

[91] *De naturalibus facultatibus* 2. 97-8 K (trans. A. J. Brock).

[92] Ibid. 211 K ὁ γὰρ δὴ τρόπος τῆς θρέψεως αὐτῶν τοιόσδε τίς ἐστι. τοῦ συνεχοῦς ἑαυτῷ σώματος, οἱόνπερ τὸ ἁπλοῦν ἀγγεῖον Ἐρασίστρατος ὑποτίθεται. The problem may, as Solmsen (1961), 191, suggests in a different context, be related to the nature of the nervous *lumina*. If so, then Galen is certainly biasing the discussion towards element theory.

[93] Lonie (1964), 441 n. 49, suggests that this was a point left open by Erasistratus himself. This is pretty doubtful in my view. None the less, some (e.g. Singer and Underwood (1962), 49) have exploited the ambiguity in order to argue that Erasistratus too was some kind of atomist. That the idea of the τριπλοκία need have nothing to do with a particulate theory of matter has been stressed most recently by Furley (Furley and Wilkie (1984), 38).

[94] 2. 95 K.

Galen's conviction that Asclepiades had a place for void in his physics. Galen seems to assume that a corpuscular theory will inevitably presuppose void, although he does not say so explicitly here. (We have already seen him playing fast and loose with other such theories.) Now, this group of Erasistrateans who posit a theory which 'arrives at Asclepiades by the garden gate'[95] correspond directly, in my view, to the 'refutationists' whom we detected in the *De usu respirationis*. Furthermore, these opponents of Erasistratus' explanation of suffocation put forward arguments which are similar in effect to those of the 'Asclepiadeans' with whom they are loosely connected by Galen at *De usu respirationis* 4. 475 K.

To summarize briefly: the refuters of Erasistratus in the *De usu respirationis* claim that when the breath is held, PTKA cannot operate because the contents of the lungs could just as easily expand into whatever space is available. The possibility of either a disseminate void or an elastic continuum, they hold, effectively banishes PTKA from explanations of respiration generally, and the role of the heart in particular. These physiological phenomena require a different kind of explanation. This is exactly what the Asclepiadeans later in the same passage seem to be offering, with their substitution of 'shaking' for cardiac expansion and contraction. Second, the suggestion of a type of disseminate void which invalidates the appeal to PTKA parallels the position of the second group of Erasistrateans in the *De naturalibus facultatibus* (2.98 K), whose elements are similar to those of Asclepiades. The language of disseminate void used at *De usu respirationis* 4. 474 K has a parallel at *De morborum causis* 7. 2 K, where Asclepiades is certainly in Galen's mind.

If Asclepiades did react in some way against Erasistratus, and perhaps even side with a group of renegade Erasistrateans, why, and how? On the basis of Galen's testimony, an obvious point of departure for Asclepiades will have been the question of void. Yet if the arguments against PTKA also involve continuum theory we may have to look a little further.

---

[95] Ibid. 98 K.

At *De naturalibus facultatibus* 2. 99 K Galen expressly says that Erasistratus denied the possibility of a 'perceptible, large void':

ὡς αὐτὸς ὁ Ἐρασίστρατος ὁμολογεῖ διαρρήδην, οὐ περὶ τοῦ τοιούτου κενοῦ φάσκων ἑκάστοτε ποιεῖσθαι τὸν λόγον, ὃ κατὰ βραχὺ παρέσπαρται τοῖς σώμασιν, ἀλλὰ περὶ τοῦ σαφοῦς καὶ αἰσθητοῦ καὶ ἀθρόου καὶ μεγάλου καὶ ἐναργοῦς καὶ ὅπως ἂν ἄλλως ὀνομάζειν ἐθέλῃς.

This Erasistratus expressly acknowledges, for he states that it is not a vacuum such as this, interspersed in small portions among the corpuscles, that his various treatises deal with, but a vacuum which is clear, perceptible, complete in itself, large in size, evident, or however else one cares to term it.

It is the impossibility of this kind of void which, of course, informs PTKA, and Galen goes on to say that disseminate void exists only 'in theory'. Whether this was Erasistratus' own view or not, is not made clear.[96] The evidence from the *De usu respirationis* which I have discussed above suggests that it is not likely; there, the opponents of Erasistratus claimed that he did not mention it at all.[97] But this latest passage on its own gives a different impression. The implication is that a theoretical void can have no active δύναμις: if one's theory can accommodate it, then it is not παρὰ φύσιν. This is a view which finds a certain amount of sympathy with Galen, if only because he can use it to fuel his long-standing hostility to the 'theoretical', especially when the theory belongs to someone else.

We can tell from Asclepiades' criticism of PTKA that he found a physical theory of the type which may have been advocated by Erasistratus unfit to account for the phenomena, just as did the ἀντιλέγοντες in the *De usu respirationis*. While an elastic

[96] It is not clear whether the distinction between large and small void was actually made by Erasistratus in these terms, or merely known to him. Cf. Anonymus Londinensis 26. 48; 27. 6ff., 28ff., 38ff. The Anonymus uses the vocabulary of large-scale void associated with Strato, but there is no sign of the corresponding concept of small void. The terminology could, of course, belong to the Anonymus rather than Erasistratus.

[97] Gatzemeier (1970), 94–7, maintains that Strato was a continuum theorist, and that Simplicius' evidence to the contrary is based on a misunderstanding. At the other end of the spectrum, Gottschalk (1965), argues for a fairly straight line from Strato to Erasistratus to Hero, with each of them positing what Furley (1989) calls microvoid.

continuum would have solved the problem, some form of disseminate void is certainly implied by Galen's talk of the 'elements of
Asclepiades' at *De naturalibus facultatibus* 2. 99 K. Granted that
this particular type of disseminate void did away with the need
for PTKA altogether, we must assume that if Erasistratus had
posited disseminate void in his own theory, either he did not
foresee the objections we have examined, or his void was different in some way from that of Asclepiades. If the refuters had
been wrong about that, then surely even Galen's Erasistrateans
would have picked it up. I can offer no firm conclusion about
whether or not Erasistratus was a continuum theorist; scholars
are still divided over Strato's position on this matter. And this is
as far as I wish to take the problematic evidence for Erasistratus'
concern with issues raised by continuum theorists in the Academy,
not to mention those in the Lyceum. (It would be very interesting
to investigate the possibility that his physiology may have
represented some kind of synthesis between the two schools.)

Does this mean that Asclepiades assigned a power to theoretical, rather than large, void, whose disseminate nature he tried to
convey with the term λεπτομερές? This is what Galen is trying
to suggest. On this interpretation, PTLP, would be a kind of *horror
vacui* theory operating at the level of individual particles. Superficially it would seem to put Asclepiades closer to Strato than
Erasistratus. It might also make sense of the fact that while
Galen never links Erasistratus with Strato, he does quote Athenaeus' opinion that Asclepiades' explanation of ῥῖγος was related
to Strato and Heraclides.[98] But how could λεπτομερές be used to
convey the sense of void like this? How would this explain why
Asclepiades was apparently so keen to speak of PTLP as involving
a φορά or a ῥύσις rather than ὁλκή or ἀκολουθία?[99] If the

---

[98] *De tremore et rigore et palpitatione* 7. 615–16 K. The relation is weak. This passage is
frequently cited by those who see Heraclides behind the Asclepiadean corpuscular
theory. All that Galen says is that Athenaeus mentioned only the discussions of ῥῖγος
offered by Asclepiades, Heraclides, and Strato when he should have mentioned other discussions which are far more plausible. This is hardly enough to link the three earlier
theorists.

[99] Strato apparently used magnetic attraction in one of his proofs of small-scale void:
fr. 61 Wehrli = Simplicius, *In Physicorum*, p. 652. 18–25; fr. 62 Wehrli = Simplicius, *In
Physicorum*, p. 663. 2–8. In the latter passage Simplicius notes that Strato held this proof

λεπτομερές had some active faculty, then it is difficult to explain why the language of attraction was not used.[100]

More important evidence comes from Galen and his distinction between the type of attraction represented by PTKA, and attraction 'to what is appropriate'. Although he does not say it in so many words, Galen himself seems to regard PTLP as belonging to the second category along with magnetism, and what we now call deliquescence. Both these phenomena are discussed in contexts where PTLP is under attack. Let us consider deliquescence first. At *De naturalibus facultatibus* 2. 55–6 K Galen sets out to refute PTLP in the following way. Sometimes unscrupulous farmers wish to make their corn stocks look more substantial than they really are. They fool their gullible customers by placing large jars of water with the corn in their carts as they are going to market. The corn, Galen argues, has the power to attract moisture from the water in the jars. Proof of this can be had from the fact that far more water will be found to have vanished from the jars than would have been the case if the water had simply evaporated. Yet, he continues, PTLP would entail that the water's motion should be more towards the surrounding air than towards the corn (because air is more fine, more λεπτομερές, than corn).

Now magnetism. Galen does not, as I have said, give Asclepiades' explanation of this phenomenon; indeed, he claims that there was no such explanation at all on offer. I shall have more to say on this; for the moment, suffice it to say that the manner of Galen's discussion of the question of magnetism, and how it bears on Epicurus and Asclepiades, still leaves PTLP firmly in the class of the 'attraction towards what is appropriate' type of theory. With this in the way of background, I should like to consider the few cases where we can see PTLP in action.

to be a poor one; the phenomena of attraction do not necessitate void. Asclepiades seems to have had similar doubts, according to Galen's account in the *De naturalibus facultatibus*.

[100] When Galen attacks Asclepiades at *De placitis Hippocratis et Platonis* 5. 640 K for 'using the term attraction (ὁλκή) instead of movement (φορά)', he is merely accusing him of pretending to have accounted for something which in Galen's view he had not explained.

THE PULSE

Asclepiades followed many ancient doctors in explaining the action of the blood-vascular system ultimately in terms of the respiratory system. For him, the arteries contained pneuma—as they did for Erasistratus[101]—pneuma which has its origin in inspired air. But there are no texts which explicitly define the nature of the link between respiration and the pulse, so I shall treat both separately, as similar aspects of the same problem. Wellmann was the first this century to suggest that Asclepiades believed that the arteries contained blood as well as pneuma. He, and those who have followed him, seem to be wrong.[102] So what was the case? Galen regarded Asclepiades' doctrine on the pulse as perverse. At *De differentiis pulsuum* 8. 747-8 K we are told that while Herophilus believed that the arteries were in a natural state in diastole, Asclepiades maintained the direct opposite. According to him, the arteries dilate when pneuma 'flows into them' because of the λεπτομερές which they contain.

ἐὰν γὰρ ἀκριβῶς ἔπηται τοῖς Ἡροφίλου δόγμασιν, ἡ συστολὴ μὲν ἐνέργεια τῶν ἀρτηριῶν ἐστιν, ἡ διαστολὴ δὲ εἰς τὴν οἰκείαν τε καὶ φυσικὴν κατάστασιν τοῦ σώματος αὐτῶν ἐπάνοδος. βούλεται γάρ, ὥσπερ ἐπὶ τῶν τεθνεώτων ὁρᾶται διεστὼς ὁ χιτὼν τῆς ἀρτηρίας, οὕτω κἀπὶ τῶν ζώντων ὅσον ἐφ᾽ ἑαυτῷ διεστάναι, τοὐναντίον Ἀσκληπιάδου δοξάζοντος· οἴεται γὰρ ὁ ἀνὴρ οὗτος καὶ τὴν καρδίαν καὶ τὰς ἀρτηρίας διαστέλλεσθαι πληρουμένας πνεύματος, εἰσρέοντος αὐταῖς διὰ λεπτομέρειαν, ἣν ἐντὸς ἑαυτῶν ἔχουσιν, ὅταν δὲ πληρωθεισῶν εἰς τὸ ἔμπροσθεν οὐκέτι ῥέῃ, καταπίπτειν αὖθις εἰς τὴν ἔμπροσθεν ὑπάρχουσαν ἑαυταῖς κατάστασιν φύσει τὸν χιτῶνα.

---

[101] Erasistratus: see e.g. *An in arteriis* 4. 705-7 K; *De locis affectis* 8. 324-5 K.

[102] Wellmann (1908), 685 n. 5. He was followed by Lonie (1965), 129 n. 1, and von Staden (1975), 182; see also Pigeaud (1981*a*), 187-8. The evidence cited by Wellmann and Lonie is quite inconclusive. (Caelius, *De morbis acutis* 1. 124, 2. 180, and Galen, *De usu partium* 3. 466ff. K, do not point either way; εὐτραφής, applied at 3. 467 K to the arteries, certainly need not mean that they were nourished by any blood they might have carried. Asclepiades is arguing here that there is no difference between the nature—the structure—of veins and arteries, not that there is no difference between what they carry. Galen presents this as part of his argument against Asclepiades' opposition to teleology.)

In strict conformity with the doctrines of Herophilus, contraction (systole) is an activity of the arteries, while dilation (diastole) represents a return to the proper natural state of their body. He means to say that just as the arterial coat is in diastole in corpses, so too it is, as far as it can be, in living creatures. Asclepiades holds the opposite view. This character holds that the heart and arteries are dilated when they are filled with pneuma which flows into them because of the λεπτομέρεια which they contain. When they are full, and the influx ceases, the coat contracts into its former, natural state.

What is this λεπτομέρεια? Asclepiades could have said that the arteries contained void if he had wanted to. And Galen presumably would have been delighted to agree. But what seems to happen is that when the arteries are full and can accommodate no more pneuma their coats collapse, and they return to their former natural state in systole. This report is repeated in the same treatise at 714 K. At 646 K there is a hint that Asclepiades used the ideas suggested by PTLP to describe different types of pulse; here a 'violent' pulse is explained in terms of πλῆθος καὶ λεπτότης πνεύματος, the implication being that the greater the extent of λεπτομέρεια the greater the amount of pneuma which flows in.[103] This is what makes the arteries shake (and lungs too, as we shall see soon). This is not to say, with Pigeaud,[104] that Asclepiades saw no regularity in the pulse. He used it as an important diagnostic tool much as other doctors did.

It is important to stress once again that the pneuma flows into the arteries and thus expands them. Galen puts it well: there is an important difference between 'dilation as a result of being filled' (τὸ πληρούμενον διαστέλλεσθαι) and 'being filled as a result of dilation' (τὸ διαστελλόμενον πληροῦσθαι). Wineskins and bags are expanded because they are filled; bellows are filled because they are expanded.[105] And here is the difference

---

[103] *De differentiis pulsuum* 8. 714 K ὁ μὲν γὰρ Ἐρασίστρατος ἐρεῖ τὸν σφυγμὸν εἶναι κίνησιν ἀρτηριῶν κατὰ διαστολὴν καὶ συστολὴν ὑπὸ ζωτικῆς τε καὶ ψυχικῆς δυνάμεως γινομένην, ἐπιπληρώσεως ἕνεκεν ἀρτηριῶν, ἐχουσῶν ἐν αὐταῖς πνεῦμα ζωτικόν. ὁ δ’ Ἀσκληπιάδης κίνησιν ἀρτηριῶν κατὰ διαστολὴν καὶ συστολήν, πληρουμένων μὲν πνεύματος τῇ πρὸς τὸ λεπτομερὲς φορᾷ, κενουμένων δὲ τῇ καταπτώσει τοῦ χιτῶνος αὐτῶν. Cf. *De differentiis pulsuum* 8. 645-6 K.

[104] Pigeaud (1981 a), 188.   [105] *An in arteriis* 4. 731 K.

between PTLP and Galen's own explanations of attraction. PTLP is in the first category above, while Galen, positing specific attractive faculties, say, for the coats of the arteries, is in the second. This, surely, is what is behind Galen's insistence in the *De naturalibus facultatibus* that Asclepiades 'denied attraction'. The exact force of φορά and ῥεῖν needs further examination. If something moves towards something else, and is not attracted (i.e. pulled) or self-moving, then it must be pushed. The presence of λεπτομέρεια in the arteries is a condition for the entrance of pneuma, but in itself exercises no attraction. It might seem reasonable after all to draw an analogy between PTLP and the best-known ancient theory of 'pushes', Plato's theory of περίωσις in the *Timaeus*. I have already noted Erasistratean involvement in this Platonic debate: Plato's account of respiration at *Timaeus* 79 A 5–C 7 bears a number of resemblances to that of Asclepiades. περίωσις as Plato describes it can only work in a strict continuum, where something is pushed into one space to make room for something else. Whether or not we are prepared at this stage to see a connection between PTLP and Plato's theory, the idea that PTLP might be part of a continuum theory at least deserves serious consideration.

RESPIRATION

A brilliant article by Fritzsche set the scene for twentieth-century work on Asclepiades' account of respiration, even though Asclepiades was by no means Fritzsche's main concern.[106] I do not ultimately agree with his thesis about Epicurus and Asclepiades, but what I have to say owes a considerable debt to his analysis of the connections between theories of magnetism and respiration.

Asclepiades' explanation of respiration somehow found its way into the doxographical tradition, but it is not discussed at all

---

[106] Fritzsche (1902); the pre-Asclepiadean primary sources are conveniently assembled by Wellmann (1901), 82–5. Once again, Harig (1983) provides a sensible over-view. Harig's final conclusions are not far from my own.

by Galen[107]—surprising perhaps. It survives in three broadly similar versions, all originating in the *Placita* of Aetius. The version I translate below comes from ps.-Plutarch:

Ἀσκληπιάδης τὸν μὲν πνεύμονα χώνης δίκην συνίστησιν, αἰτίαν δὲ τῆς ἀναπνοῆς τὴν ἐν τῷ θώρακι λεπτομέρειαν ὑποτίθεται, πρὸς ἣν τὸν ἔξωθεν ἀέρα ῥεῖν τε καὶ φέρεσθαι παχυμερῆ ὄντα, πάλιν δ᾽ ἀπωθεῖσθαι, μηκέτι τοῦ θώρακος οἵου τ᾽ ὄντος μήτ᾽ ἐπεισδέχεσθαι μήθ᾽ ὑποστέγειν· ὑπολειπομένου τινὸς ἐν τῷ θώρακι λεπτομεροῦς ἀεὶ βραχέος (οὐ γὰρ ἅπαν ἐκκρίνεται), πρὸς τοῦτο πάλιν τὸ εἴσω ὑπομένον τὴν βαρύτητα τοῦ ἐκτὸς ἀντεπεισφέρεσθαι. ταῦτα δὴ ταῖς σικύαις παρεικάζει· τὴν δὲ κατὰ προαίρεσιν ἀναπνοὴν γίνεσθαί φησι, συναγομένων τῶν ἐν τῷ πνεύμονι λεπτοτάτων πόρων καὶ τῶν βραγχίων στενουμένων· τῇ γὰρ ἡμετέρᾳ ταῦθ᾽ ὑπακούει προαιρέσει.

Asclepiades likens the lung to a funnel. He supposes that the cause of respiration is the fineness in the chest, towards which thick air flows from outside. It is pushed back again when the chest is unable to receive more or contain it. A small amount of fineness always remains in the chest (for it is not all excreted) and it is towards this which remains inside that the weight from outside is borne back in again. He likens the process to what happens with cupping-glasses. He says that voluntary respiration takes place when the finest pores in the lung are gathered together and the bronchial passages are narrowed. For these things obey our will.[108]

The ps.-Galenic account lacks the note that Asclepiades drew some kind of analogy between the lung and a funnel (χώνη).[109] But they all agree that Asclepiades held breathing to be brought about by the presence of λεπτομέρεια in the abdominal cavity 'towards which dense air flows from outside, and whence it is pushed when the chest can contain no more'. Once again, no

---

[107] In the *De usu respirationis* (4. 475 K) Galen regards the theory as beneath his notice.
[108] Ps.-Plutarch, *Placita* 903 E-F. For the other versions, ps.-Galen, *De historia philosophica* 103; Aetius Arabus 4. 2. 1-2. 8 Daiber.
[109] The analogy may just relate to shapes which naturally gather or collect material, without any intrinsic power of attraction. At Aristophanes, *Thesmophoriazusae* 18, the word is used of the ear (and its ability to gather sounds) (cf. Plato, *Republic* 3, 411 A 5 ff., which supports Reiske's emendation of the Aristophanes text). For a rather different sense of χώνη see Empedocles 31 B 84 11. 10-12 DK; and on this, Guthrie (1965), 212 n. 1. It is just possible that the analogy may, however, look back to the ἐγκύρτια of Plato's analogy in the *Timaeus*.

sign of PTKA. As in the case of the arteries, the air flows in from a
level of higher to a level of lower density; we may assume that
the rising of the chest on inhalation is caused by this influx, and
that the muscles of the chest have no role in inhalation. Respiration, the account continues, carries on because there is always
some λεπτομερές remaining in the chest. The λεπτομερές is
characterized as a gathering of the finest pores in the lung. This
is a traditional idea: Erasistratus was only one ancient doctor
who likened the lung to a sponge. An explanation which appeals
to the existence of density gradients, and evidently not to the
phenomenon of *horror vacui*, looks as if it might well have a place
somewhere for the kind of void in which the atomists believed.

Slightly different types of density gradients were also
employed in Erasistratean physiology; at *De usu respirationis* 4.
496 K Galen quotes an Erasistratean explanation of how people
suffocate in certain surroundings. Erasistratus, claims Galen,
holds that newly plastered houses are unhealthy places to live in
because the air they contain is λεπτός, and unable to hold the
pneuma in the arteries. It flows out, and suffocation ensues.[111] It
is notable, however, that this type of case seems not to have
been connected in any direct way with PTKA. The dynamics of
that process cannot be the same as those in PTLP. Under the
terms of the Asclepiadean theory, the λεπτομερές does not
seem to have any attractive faculty (as the threat of void would
have in PTKA); rather it seems merely to represent a space available to accommodate an influx of matter.

Asclepiades apparently illustrated the process of breathing
through a comparison with cupping-glasses. This, it might be
said, is a highly significant analogy; cupping-glasses would
appear, prima facie at least, to provide a good case of *horror vacui*
at work. But in fact this is of little help to us. Cupping-glasses
were used to illustrate just about any theory of attraction, simply
because they offered to doctors a good and convenient example
of the phenomenon.[112] Their exact mode of operation was prob-

---

[110] See *De locis affectis* 8. 324–5 K, and PW s.v. 'Schwamm' for other doctors.
[111] The same explanation of suffocation, this time couched in terms of PTLP, occurs at
Cassius, *Problemata* 78.   [112] *Pace* Fritzsche (1902), 384.

ably never fully understood in antiquity. All the present reference tells us is that Asclepiades had an explanation of how cupping-glasses worked, and that it was probably in terms of PTLP.

There seems little doubt that the Asclepiadean explanation of respiration goes back to Plato. This was Fritzsche's view. That περίωσις remained an influential doctrine seems clear from Galen's detailed and extended refutation of it at *De Placitis Hippocratis et Platonis* 5. 707–19 K.[113] There is good reason to believe, as we have seen, that Erasistratus examined and adapted the Platonic theory himself; Galen's commentary on the *Timaeus* suggests that there was a controversy in the Academy after Plato over the interpretation of the theory, and that Erasistratus was somehow involved in it. We might imagine that PTKA represents the outcome of Erasistratus' criticisms of this part of the Academy's physical doctrines as much as his assimilation, or perhaps just awareness of doctrines prevailing in the Lyceum. Asclepiades' account in turn could be an attempt to undo some of the Erasistratean modifications. Asclepiades did, after all, retain many Erasistratean elements in his own physiology, most notably the idea that the arteries contain pneuma, and that internal pneuma (including 'psychic' pneuma) is derived from the air we breathe.[114]

THE MOVEMENT OF CORPUSCLES IN THE BODY

PTLP was invoked in some cases to explain the gathering of corpuscles which leads to the blockage of pores in the body. This

---

[113] And, most important, a theory remarkably similar to the Asclepiadean occurs in the work of the Platonist Timaeus of Locri, περὶ φύσιος 63–5 = Thesleff (1961), p. 221 ἁ γὰρ σικύα καὶ τὸ ἤλεκτρον εἰκόνες ἀναπνοᾶς ἐντι. ῥεῖ γὰρ διὰ τῶ σώματος ἔξω θύραζε τὸ πνεῦμα, ἀντεπεισάγεται δὲ διὰ τὰς ἀναπνοᾶς τῷ τε στόματι καὶ ταῖς ῥισίν, εἶτα πάλιν οἷον Εὔριπος ἀντεπιφέρεται εἰς τὸ σῶμα, τὸ δὲ ἀνατείνεται καττὰς ἐκροάς. The relations of this explanation to the Platonic account have been examined in detail by Harder, PW suppl. viA, s.v. 'Timaios' (4), though Harder has nothing to say about Asclepiades.

[114] For Erasistratus see *De locis affectis* 8. 314–16 K. Asclepiades held that we breathe 'for the sake of the generation of the soul': see *De usu respirationis* 4. 471 K. Calcidius, *In Timaeum*, p. 229 W.

sheds some light on the present problem, but I shall go into this in more detail when I discuss the concept of blockage.

Swelling and inflammation are linked by Galen in the *De methodo medendi* to a 'flowing' of contiguous material 'towards a part which is heated'.[115] The passage is significant because while Galen believes that the humours in such cases are *attracted* to the heated part, he goes out of his way to stress that Asclepiades would have spoken of a 'flow' rather than an 'attraction'. Compare *On Medical Experience* 28, where Walzer's translation of Galen's exposition of the Asclepiadean aetiology of phrenitis runs as follows:

For you [Asclepiades] say: 'burning fever inflames the cerebral membrane, and it results from this that the atoms make their way to the *finely divided thing*, or those of them which do so become extremely fast and violent in motion all at once; this is followed by a stoppage of the atoms in the pores which causes the disease known as phrenitis. Thereupon what lies beneath the cartilages spreads upwards, being attracted[116] by the finely divided thing (τὸ λεπτομερές).'

In both the example from the *De methodo medendi* and that above, the presence of fever in the affected part of the body is a condition for the flow of material to begin. Caelius Aurelianus' juxtaposition of 'spiritus' (his translation of λεπτομερές) and 'fervor' at *De morbis acutis* 1. 119 confirms our suspicions that heat for Asclepiades is one factor at least which is capable of bringing about rarefication.

Cassius the Iatrosophist, a strange figure in many senses of the word but at least not hostile to Asclepiades, is aware of a similar kind of movement, which he attributes to Asclepiades. At *Problemata* 40 he quotes from Asclepiades' lost treatise on wounds:

διὰ τί προσπαίσματος γινομένου, τὰ μὲν σύνεγγυς οὐ συμπάσχει, τὰ δὲ πολὺ ἀφεστῶτα συνδιατίθεται, οἷον βουβῶνες τοῖς ἐν ποσὶ προσπαί-

---

[115] *De methodo medendi* 10. 878 K τὸ μὲν οὖν ἐπὶ τὸ θερμαινόμενον ἤτοι γ' ἕλκεσθαι τοὺς πλησιάζοντας χυμούς, ὡς ἡμεῖς φαμεν, ἢ ὡς Ἀσκληπιάδης ἐνόμιζε, ῥεῖν.

[116] The use of the word 'attracted' here seems very odd; the only other place where it is attested in this kind of context is Cassius, *Problemata* 78, where Asclepiades is not named.

σμασιν; ὁ οὖν Ἀσκληπιάδης ἐν τῷ περὶ ἑλκῶν φησιν ὅτι πρὸς τὰ
πληττόμενα κατ᾽ ἀρχὰς ἡ ὕλη φέρεται, καὶ φερομένης αὐτῆς, ὅσον μὲν
δύναται ὑποδέξασθαι τὰ πεπονθότα μέρη, αὕτη χωρεῖ εἰς αὐτά.
πληρωθέντων δὲ τούτων, καὶ μὴ δυναμένων ἐπιδέξασθαι ἕτερον
πλῆθος, ἡ φερομένη ὕλη ἐκρέουσα καὶ μὴ ὑποδεχθεῖσα ὑπὸ τῶν
μερῶν, ἐφ᾽ ἃ ἠνέχθη, εἶτα φερομένη, ἐὰν ἐπιτύχῃ κοίλων τόπων, μένει
εἰς αὐτούς, ὥσπερ καὶ ἐπὶ τῶν ὑδάτων ἔχει. ταῦτα γὰρ ἕως μὲν ἐπ᾽ ἰσο-
πέδῳ φέρεται, ὁμαλῇ τῇ κινήσει χρῆται· τυχόντα δὲ κοίλων τόπων,
μένει εἰς αὐτούς. ταὐτὸν οὖν συμβαίνει καὶ ἐπὶ τῆς φερομένης ὕλης ἐπὶ
τὰ πληγέντα. ὅσην γὰρ ὑποδέξασθαι δύναται, αὕτη χωρεῖ εἰς αὐτά· ἡ
δὲ λοιπὴ εἰς κοῖλα, καὶ μᾶλλον εἰς ἀραιοπόρους ἐμπίπτει τοὺς
βουβῶνας, καὶ διογκοῖ τούτους. ἔστι μὲν οὖν καὶ αὕτη πιθανὴ ἡ ἀπο-
λογία.

When someone stubs one's foot, why is it that there is no local affection, while parts which are further removed are affected? For example,
people who stub their feet get swollen glands. Asclepiades in his *On
Wounds* says that material is first carried to the parts which have been
struck. The material is carried there and approaches the affected parts
in proportion to their ability to accept it. When they are full, and can
take in no more, the matter carried there flows out, and since it has not
been accepted by the parts to which it was borne, is then carried on. If
it reaches internal cavities it stays in them, as happens in the case of
water, which is borne along with a uniform motion as long as the
ground is level, but on coming across depressions in the landscape
remains in them. The same thing happens to the material which is
borne towards the parts which have been struck. As much material as
can be accommodated moves into them, but the rest goes into hollow
parts, and especially into the glands with their fine pores, making them
swell. This is a convincing account.

There are many important things to note here. Most significant
is the comparison of movement of material in the body with that
of surface water. Water moves from higher to lower ground
unless there is a depression, in which case it runs into that and
stays there. The initial movement of material to the affected part
is seen as a movement εἰς ἀραιοπόρους, reflecting a gradient
from higher to lower density. The areas of lower density are
analogous to the geographical depressions. We might characterize this as a ῥύσις εἰς ἀραιοπόρους, just as we might call the

example at Galen *De methodo medendi* 10. 878 K a ῥύσις ἐπὶ τὸ
θερμαινόμενον.[117] Cassius seems to have used PTLP or something
very like it in several places where Asclepiades is not named.
Problemata 75, for example, seeks to answer the question why
people with headaches become dim-sighted and weep with an
appeal to a φορὰ ἐπὶ τὰ πεπονθότα. At *Problemata* 78 Cassius
gives a hint that λεπτομέρεια is to be found in the affected parts.
Again, heat, which brings about λεπτομέρεια in the belly, is
behind his rather elaborate explanation of why 'in cases of
cholera the extremities go into spasm and cool, while the pulse
is weak'.

URINARY SECRETION

Unfortunately, we do not know exactly how Erasistratus
explained how urine comes to end up in the bladder, even though
Galen relates the Erasistratean account to the Asclepiadean
explanation.[118] We do know that Asclepiades employed PTLP in
his account, because Galen attacks it at length in *De naturalibus
facultatibus* 2. 32 K, a passage which I have already mentioned.[119]
The bladder here represents the λεπτομερές, just as the lungs did
in the case of respiration. The analogy between bladder and
sponge is an old one; it also helps us to link the different examples
of PTLP at work. The process envisaged by Asclepiades involves
first of all the dissolution of water into 'vapours'. By 'vapours'
Galen is evidently pointing to what other sources call pneuma.[120]
Since pneuma is made up from the finest corpuscles, we may
imagine this alteration from liquid to vaporous states as being
the result of fragmentation of μέρη ὑγροῦ.[121] The verb which

[117] As Wellmann in fact did, although in Wellmann (1908), 700 n. 5, he gives the
impression that the expression is to be found in the text.
[118] See *De naturalibus facultatibus* 2. 187 K.
[119] See above, n. 35.
[120] And what Sextus Empiricus might call the 'parts of pneuma'. Cf. Galen, *De usu
partium* 3. 647, and *De differentiis febrium* 7. 277–8 K. Also Fischer (1982), 51 n. 7.
[121] The same kind of dissolution is attributed to Asclepiades at *De elementis* 1. 490 K.
This seems further proof that Asclepiades did not envisage two different levels of particle
in his physics.

Galen uses to describe this dissolution, ἀναλύω, carries with it an obvious echo of *solvere*.[122] The reduced size of the particles gives them a greater mobility and penetrating power; they flow towards an area of lower density represented by the bladder. Once there, they regroup to form liquid.[123]

I have suggested that PTLP grew out of PTKA; but also that PTLP shows a number of features independent of its Erasistratean antecedent. I also suggested that Asclepiades may have been a continuum theorist. But in its lack of a hydraulically significant type of void, PTLP may yet be demonstrating its atomistic parentage. This would certainly explain Galen's testimony, and his constant insistence, that Epicurus and Asclepiades had at least superficially similar conceptions of vacuum.

That, of course, is a far cry from saying that Asclepiades was himself an atomist. Yet the question of Epicurus still hovers in the background, and one modern argument deserves particular attention. Lück (1932) took up a line suggested by Fritzsche (1902) and maintained that PTLP could be related to the Lucretian explanation of magnetic attraction in the *De rerum natura*.[124] He pointed out that the Lucretian account of the phenomenon is not at all the same as the only surviving account attributed to Epicurus, which, coincidentally or not, is the one in Galen's *De naturalibus facultatibus*. A different Epicurean tradition was postulated by Lück to explain this divergence. Here is the central part of Lucretius' account of the lodestone:

> Principio fluere e lapide hoc permulta necessest
> semina, sive aestum qui discutit aera plagis
> inter qui lapidem ferrumque est cumque locatus.
> hoc ubi inanitur spatium multusque vacefit
> in medio locus, extemplo primordia ferri
> in vacuum prolapsa cadunt coniuncta, fit utque
> annulus ipse sequatur eatque ita corpore toto.

---

[122] An idea I shall develop in the next chapter.

[123] The idea of fine particles turning into liquid occurs also in the Asclepiadean aetiology of dropsy (Caelius, *De morbis acutis* 1. 108), and in his account of bilious secretion (*De naturalibus facultatibus* 2. 39-40, 66-7 K).

[124] *De rerum natura* 6. 998-1041.

Firstly, a great many seeds must flow out from this stone, or some kind of blast which with its blows scatters the air which lies between the stone and the iron. When this space is being evacuated, and a large vacuum is being created in the middle, then at once the atoms of the iron are borne into the empty space, and fall in together. Then the [iron] ring itself follows on, and proceeds this way in its entirety.[125]

Several similarities between this account and Asclepiadean PTLP are apparent at once:

    1. An area of void (Lucretius) or low density, λεπτομερές (Asclepiades), is created by emanations from the magnet for Lucretius, and by the heat in fever for Asclepiades in at least one case.

    2. The surrounding matter flows in towards what is void or λεπτομερές. This is partly because void and λεπτομέρεια offer less resistance than what contains more matter.

The Epicurean explanation reproduced by Galen is quite different, though more what we might expect from an atomist. That account is based on the idea that atoms which flow out from the lodestone have some kind of physical affinity with those in the piece of iron, which allows them to link up with one another.[126] The Lucretian account has nothing like this: in fact the process is likened a little further on to the way a ship is pushed onwards on its course by the wind.[127] There is, of course, no problem in imagining that the Epicureans had a variety of explanations of magnetism. Yet the links between my reconstruction of PTLP and the Lucretian account of magnetic attraction are too significant to ignore. We may imagine either that Lucretius has taken on a non-atomistic model—from Asclepiades himself? (I doubt it); or that we have here some sign of what Galen calls Asclepiades' Epicurean inheritance.

    Even though we know next to nothing about Asclepiades' intellectual background before he came to Rome, the tradition of Greek Epicureanism in the second and first centuries was strong. Once in Rome, Asclepiades continued to write in Greek—that is

---

[125] *De rerum natura* 1002–8.
[126] *De naturalibus facultatibus* 2. 45 K.
[127] *De rerum natura* 6. 1030–3.

almost certain—and was regarded as a Greek. But his greatest successes were among Romans, at a time when Epicurean philosophy had firmly established itself, for better or worse, among the Republican intelligentsia. He should thus be seen against the background of an age which produced such prominent Greek and Roman Epicureans as Amafinius, Zeno, Rabirius, Phaedrus, Varus, Torquatus, Alcaeus, Catius, Lucullus, and Lucretius. And that is without all those Epicureans at Herculaneum. Cicero himself, their bitter opponent, could number many Epicureans among his friends; he had to admit that there was hardly anyone who had *not* been affected by them in some way.

Consequently, when Pigeaud makes much of Asclepiades' use of 'Epicurean terminology', we must ask ourselves just how much that proves.[128] Yes, it is true that many of our witnesses attribute to him phrases like ἐν λόγῳ θεωρητόν in connection with the corpuscles. But this kind of language is not specifically diagnostic of Epicureanism; it is also used in connection with the theories of Herophilus and Erasistratus.[129]

The background to Asclepiades' corpuscular hypothesis is steadily coming into view. Thus far, we have a system based on the idea of fragile corpuscles, not at all like Epicurean atoms, moving about in what may after all have been an Epicurean void. It is not necessary to reject Galen's testimony about Asclepiades and the void, even if we can question the motives behind his assimilation of Asclepiades and Epicurus. If this is correct, then the theory is unique in antiquity. It offers an unparalleled glimpse of just where the tensions between atomism and continuum theory could lead. Asclepiades may well have reacted against PTKA precisely because of Erasistratus' line on the power of void, while at the same time preserving a basically Erasistratean model of how the body is composed. That is an idea I shall develop presently. First, however, we need to look more closely at how this strange theory operated in practice.

---

[128] Pigeaud (1981*a*), 171.

[129] At *De simplicium medicamentorum temperamentis ac facultatibus* 11. 783 K Galen says that all those with corpuscular theories see visible bodies as συγκρίματα. This is not to say they were closet Epicureans.

3
The ὄγκοι in Action

Asclepiades' theory was quite spectacularly speculative. Health was nothing but the free, 'balanced' movement of fluids within the body, an 'equilibrium of corpuscles and pores', as Galen says.[1] The ancient witnesses suggest that every detail of Asclepiades' pathology could be subordinated to the principle that a disturbance of the movement of corpuscles causes some kind of dysfunction.

κατὰ δ' Ἐρασίστρατον καὶ Ἀσκληπιάδην, ὡς ἐπίπαν μίαν αἰτίαν ἐπὶ πάσης νόσου, καθ' ὃν μὲν παρέμπτωσις εἰς τὰς ἀρτηρίας τοῦ αἵματος· καθ' ὃν δὲ ἡ ἔνστασις τῶν ὄγκων ἐν τοῖς ἀραιώμασιν.

According to Erasistratus and Asclepiades there is just one cause of all disease. For Erasistratus it is the transference of blood into the arteries. For Asclepiades it is the impaction of corpuscles in pores.[2]

In ascribing illness to an imbalance, Asclepiades and Erasistratus stand in a long tradition that certainly goes back to the Hippocratics, even if at this level of generality there is little to distinguish them from other doctors with physiological theories. The Hippocratic concept of ἰσονομία comes obviously to mind here, but note too the role of balance in Plato's pathology in the *Timaeus*.[3] Health was a συμμετρία of some kind for all doctors who sought such explanations, says Galen: συμμετρία γὰρ δή τις

---

[1] συμμετρία τῶν ὄγκων καὶ πόρων at *De placitis Hippocratis et Platonis* 5. 449–50 K; cf. *De methodo medendi* 10. 268 K οὐ γὰρ ἁπλῶς ὡς Ἀσκληπιάδης ἐν συμμετρίᾳ μέν τινι πόρων τὸ ὑγιαίνειν ἡμᾶς ὑποθέμενος, ἐν ἀμετρίᾳ δὲ τὸ νοσεῖν.
[2] Ps.-Galen, *Introductio sive medicus* 14. 728–9 K. See also Celsus, *De medicina* 1, proem 15–16; Galen, *De methodo medendi* 10. 101–2 K. On the universality of the corpuscular hypothesis see the Anonymus Londinensis 39. 22–32.
[3] Esp. *Timaeus* 82 A–86 A, with Miller (1962).

ἡ ὑγίεια κατὰ πάσας ἐστι τὰς αἱρέσεις.[4] But in the case of Asclepiades' theory, a series of ostensibly radical doctrines followed on from the basic idea. Asclepiades is supposed to have rejected the analysis and treatment of disease in terms of critical days,[5] down-graded the causal role of the humours,[6] denied that gynaecology exists as a separate branch of medicine,[7] and refused to accept that there is any such thing as 'innate' bodily heat, with all the assumptions about Nature that this concept necessarily entailed.[8] Just how radical these doctrines actually were is another matter. Celsus and Pliny tend to underline Asclepiades' claims to innovation, but at the same time they suggest that his explicit claims to innovation, especially in the field of his medical practice, were at least partly rhetorical and concealed a fair degree of dependence on his predecessors.[9]

Caelius is by far our most copious source for the workings of Asclepiades' pathology. This is rather unfortunate when it comes to learning the details; Caelius and Soranus sport a lively contempt for theory, and provide very little of the kind of in-

---

[4] *De sanitate tuenda* 6. 15 K; cf. *Commentarium in Hippocratis de natura hominis* 15. 60 K (and 61 K) κατὰ πάντας ἰατρούς τε καὶ φιλοσόφους τοὺς τελείους δογματικοὺς ἡ συμμετρία τῶν στοιχείων ὑγίειαν ἐργάζεται. For similar cases of Galen drawing together this aspect of all the pathological systems he knows, see *De methodo medendi* 10. 116–17 K and *De morborum differentiis* 6. 838, 853–4 K. Consequently, those seeking Asclepiades' specific affiliations should not read too much into the occurrence of συμμετρία and πόροι together at Epicurus, fr. 36. 23. 3–5 Arrighetti, nor into ἡ ἀσυμμετρία τῶν πόρων at Theophrastus, *De igne* 42.

[5] Because the 'right moment' in treatment is the creation of the skilled doctor. See Caelius, *De morbis acutis* 1. 109; Galen, *De crisibus* 9. 735–6 K; *De diebus decretoriis*, 9. 798–9 K. Just how far he went is unclear; at *De morbis acutis* 1. 108 Caelius implies that Asclepiades retained the chronologically defined fever terminology of his predecessors. Celsus, *De medicina* 3. 4. 11–15, approves of Asclepiades when he says that no day is more or less dangerous for its being odd or even.

[6] Caelius, *De morbis acutis* 1. 112; the 'alii' mentioned at *De morbis acutis* 3. 188 who argue that the discharge in cholera is not bile, but a liquid which looks like bile, could be Asclepiadeans, to judge from the context.

[7] Soranus, *Gynaecia*, 3. 3. 5. Asclepiades is not mentioned at all in Caelius' own Latin version of Soranus' *Gynaecia* (=Drabkin (1951)).

[8] e.g. Galen, *De tremore, rigore et palpitatione* 7. 614–15 K.

[9] For the claims to innovation see Pliny, *Naturalis historia* 7. 123–4, 26. 12–20, 29. 5–6; Celsus, *De medicina* 1, proem 10–11. Celsus is keen to stress that the claims were not always justified: see e.g. *De medicina* 2. 14. 1–2 on the Asclepiadean 'discovery' of the therapeutic benefits of *frictio*. Galen, predictably enough, joins in: see *On Medical Experience* 13. 7–8. On the whole matter of claims to innovation in ancient science, and why they might have been made, see Lloyd (1987), ch. 2.

formation we might want. Soranus merely accuses Asclepiades of 'lying' about causation and the elements at *Gynaecia* 3. 4 (although the text is uncertain, and Soranus may be directing his attack on element theory at Herophilus rather than Asclepiades). Caelius generally claims to suspend judgement about the corpuscular hypothesis, but in practice his hostility is uniform. Theory, says Caelius, is admissible as long as it is regarded as purely speculative.[10] Methodist criticisms of Asclepiades are directed not so much at the details of the corpuscular hypothesis, but at the very fact that Asclepiades attached such importance to speculating about the 'nature of man' in the first place.[11] Here, I shall argue, Asclepiades ultimately chose his own tree from which to hang.

Caelius' two major works follow a fairly strict pattern in which explanations of diseases by different doctors are set out, before they are rejected and the appropriate Methodist therapy extolled.[12] I have laid out below all those diseases for which Caelius presents an Asclepiadean aetiology. (I say 'Asclepiadean' because I also include certain aetiologies attributed to the followers of Asclepiades.) Caelius calls these explanations 'definitions', a term which for him has a pejorative sense since it signifies a theory-laden approach. We should not assume that

[10] Cf. *De morbis acutis* 2. 147–8, 183, 187.

[11] Note the sceptical arguments ranged against the corpuscular hypothesis by Caelius in bk. 1 of the *De morbis acutis*; at 1. 9 he says that Asclepiades should not have explained disease 'per occultam atque dissonantem obtrusionem et quae fortasse neque esse probetur'. 1. 9–14 provides an example of an attack on Asclepiades based on supposed ambiguities in his terminology. Does Asclepiades mean 'fever' or 'being feverish'? 'a thickened fluid' or the 'thickened parts of a fluid'? (1. 10, 13.) This particular case has close affinities with a philosophical debate over the use of predicables and appellatives in causal statements (e.g. at Sextus Empiricus, *Pyrrhoniae hypotyposes* 3. 14), but Caelius is being more rhetorical than philosophical here.

[12] Smith (1979), 224, describes Caelius' system of organization as 'hamartography'; he suggests that Asclepiades adopted much the same structure in his pathological work. I cannot really see how a system of arrangement which suits the rather self-conscious scepticism of Caelius could have been advantageous to Asclepiades, who, after all, was advancing a rigorous theory of his own. Any argument that this format is traditional would also have to explain why the broadly similar arrangement of diseases in the supposedly Methodist 'Anonymus Parisinus' texts (Fuchs (1903), e.g. pp. 69–70) does not include refutations of the theories put forward. Yet the author of the Hippocratic treatise *De vetere medicina* was following a pattern not wholly dissimilar to this one, and Smith's remains an interesting suggestion.

Caelius or Soranus offer us consistent reports of Asclepiadean explanations; it is in their interest to underline discrepancies and dissonance in Asclepiades' pathology. None the less, a useful pattern does emerge. The Latin disease names, and in a few cases the Greek equivalents, belong to Caelius.

(i) *Phrenitis*: 'statio sive obtrusio in cerebri membranis' ('an impaction or blockage in the cerebral membranes'): *De morbis acutis* 1. 6; cf. Galen, *On Medical Experience* 28. 3 Walzer 'a stoppage of the atoms in the pores which causes the disease known as phrenitis'.

(ii) *Lethargia*: Asclepiades did not 'define' this disease (*De morbis acutis* 2. 5), and the explanation given by Caelius is that of the Asclepiadean-cum-Herophilean Alexander of Laodicea (Philalethes): 'subita et recens alienatio' ('sudden and recent mental alienation'): *De morbis acutis* 2. 5.[13])

(iii) *Pleuritis*: 'humoris fluor temporis parvi atque celer interiorum lateris partium' ('a brief and acute flow of liquid in the internal lateral parts'): *De morbis acutis* 2. 89.

(iv) *Pneumonia*: 'parvi temporis solutio' ('a *solutio* of brief duration'): *De morbis acutis* 2. 142.

(v) *Cardiaca passio*: 'tumor secundum cor corpusculorum coacervatione sive obtrusione effectus' ('an inflammation in the heart brought about by a piling up or blockage of corpuscles'): *De morbis acutis* 2. 163.

(vi) *Synanche*: 'humoris fluor sive humectatio faucium vel summitatis ipsarum, saepissime ex capite accidens' ('a flow of liquid, or moistening in the throat (or at least the highest parts), very often starting from the head'): *De morbis acutis* 3. 5.

(vii) *Conductio* (σπασμός): 'extentio corporum' ('stretching of the body'): *De morbis acutis* 3. 63. But note 3. 64 'alii nostrae sectae scriptores [Methodists] pro tensione [i.e. extentio] conclusionem vocaverunt, sub alio nomine id ipsum significantes' ('Other writers from our sect speak of "blockage" instead of "stretching", using another word to mean the same thing').

---

[13] The text is corrupt here; *alienatio* is supplied by the corrector of one of the early printed editions, on analogy with its use at *De morbis acutis* 2. 6.

(viii) *Hydrophobia*: Caelius merely states Asclepiades' view that, like other diseases which disorder the mind, it is located in the cerebral membranes (*De morbis acutis* 3. 112).

(ix) *Tormentum* (εἰλεός): 'tormentum est contortio extenta atque longi temporis intestinorum' ('*tormentum* is a severe and prolonged twisting of the intestines'): *De morbis acutis* 3. 139.

(x) *Cholera*: 'humoris fluor, celer ac parvi temporis ventris atque intestinorum ex concursu sive obtrusione corpusculorum, atque ut saepe contingit ex indigestione initium sumens' ('a flow of liquid, acute and brief, from the belly and intestines, caused by the confluence and impaction of the corpuscles; it often begins with indigestion'): *De morbis acutis* 3. 188.

(xi) *Defluxio* (διάρροια): 'rheumatismus sive fluor parvi temporis ultimarum partium coli atque sessionis sive longanonis ut nos appellamus, quae fit ... ex conventu sive concursu atque congressu corpusculorum' ('an acute flux or flow from the lowest parts of the colon (i.e. rectum or *longanon*) which comes from the confluence, congregation, or crowding of the corpuscles'): *De morbis acutis* 3. 220.

The diseases seem to fall into two broad categories: those whose origin lies exclusively in blockage (*phrenitis, cardiaca passio, conductio, tormentum, cholera*, and, surprisingly enough, *defluxio* (diarrhoea)) and those which seem to involve another kind of process (*lethargia, pleuritis, pneumonia, synanche*). Caelius and Soranus elsewhere add *boulima* (dropsy) and some mild fevers to this second class of affections, caused by 'disturbance'.[14] There are, needless to say, a number of problems with this simple categorization. *Lethargia* is a doubtful case, if only because the explanation supplied is not attributed to Asclepiades himself but a supposed follower (and a disaffected one at that, who later became a Herophilean). Caelius uses a number of terms for 'blockage', yet it is not immediately easy to see exactly what is behind, say, *conductio* and *tormentum*. The case with diarrhoea, *pleuritis*, and *cholera*[15] in particular points to the

---

[14] Soranus, *Gynaecia* 3. 4; Caelius, *De morbis acutis* 1. 107–8.

[15] At *De morbis acutis* 1. 107 'pleurisy' is included in the list of affections caused by

somewhat paradoxical conclusion that flux, *fluor* itself results from a blockage. Can the problems be resolved?

THE CONCEPT OF BLOCKAGE

The idea that impeding the flow of fluids in the body can cause various types of problems is a very common one in Greek medicine. So common, in fact, that its adoption by Asclepiades is ultimately quite instructive. It figures as an important part of the pathology of many Hippocratic works.[16] It seems to have been a significant part of Diocles', Praxagoras,' and Erasistratus' pathology.[17] This is to say nothing of certain marginal figures like Aegimius of Elis and Timotheus of Metapontum.[18] That the concept should be so common need not surprise us. A large number

statio, but at ibid. 2. 89, in what purports to be an excerpt from the Asclepiadean work *On Definitions*, Caelius says that the disease is a *humoris fluor*, with no mention of blockage at all.

[16] Lonie (1981*a*), 113, draws attention to the wide use of this idea in the pathology of a variety of Hippocratic treatises.

[17] Diocles, fr. 51 Wellmann (1901)=Fuchs (1894*a*), 541-2: Diocles thought epilepsy was caused by ἔμφραξις περὶ τὴν παχεῖαν ἀρτηρίαν; similarly, headache occurs when passages in the head are blocked (Fuchs (1894*a*), 543). This particular type of explanation goes back to the Hippocratic *De morbo sacro* 7, and also figures in Praxagoras (*ap.* Fuchs (1894*a*), 541). For Diocles' doctrine that pleurisy is the result of an ἔμφραξις τῶν περὶ τὰς πλευρὰς φλεβῶν see Fuchs (1903), 93. See also fr. 40 Wellmann (1901=Fuchs (1894*a*), 548; fr. 59 Wellmann (1901)=Fuchs (1894*a*), 543; fr. 63 Wellmann (1901)=Fuchs (1894*a*), 544; fr. 68 Wellmann (1901)=Fuchs (1894*a*), 553; fr. 72 Wellmann (1901)=Fuchs (1894*a*), 547; on epilepsy: Galen, *De locis affectis* 8. 185-8 K. Blockage also figures in Galen's own explanation: see *De locis affectis* 8. 173-4 K. Compare also Diocles' explanation of phrenitis: 'freneticam passionem inquit fieri tumore in corde effecto et suffocato sanguine seu calore consuetudinario, ex quo cerebrum sensum et intellectum praebet' (Vindicianus, *De semine* 44, Wellmann (1901), 234) with the Asclepiadean explanations of both phrenitis and the *cardiaca passio* given by Caelius Aurelianus (*De morbis acutis* 1. 6-15; 2. 163). On the other hand, the Anonymus Londinensis (fr. 1 Diels) claims that Diocles regarded phrenitis as an inflammation of the diaphragm.

[18] Lonie (1965), 128. 2, suggested Timotheus as a possible precursor of Asclepiades. The only mention of his pathology is by the Anonymus Londinensis 8. 11-34: a reference to a theory which stated that blockage of 'residues' in the belly (or head if we do not accept Diels's κοιλίη for κεφαλή) causes disease. Wellmann (1908), 687, suggested Aegimius of Elis mainly because he may have held a theory of *plethora* similar to that of both Erasistratus and Asclepiades (see Anonymus Londinensis 13. 21-44). In both these cases the evidence is more than a little tenuous.

of easily observable and very common morbid conditions seem to involve blockage: choking and other respiratory difficulties, constipation, some temporary forms of deafness, and 'stones' in the kidney which lead to problems with urinary retention are all cases noted by Galen and described with the language of ἔμφραξις. We have already seen that the idea of the body as a network of pores was an attractive one for ancient doctors. The adoption of 'blockage' as a universal explanation of illness is just one extension of this idea.[19]

Even in theoretical explanations of pathological phenomena, Galen uses the idea extensively at the same time as he criticizes its use by Asclepiades. In fact, there is often little to distinguish general descriptions of blockage in Galen's pathology from those specifically attributed to Asclepiades. Galen can speak about macroscopic and microscopic structures (though not necessarily 'theoretical') much as we would expect to find Asclepiades talking about those at a microscopic and 'theoretical' level.[20] We could, I suppose, make a distinction between Galen and Asclepiades along such lines as these: where Galen can argue that blockage can be caused by an unnatural swelling or foreign body, together with the contraction of the pores,[21] Asclepiades' blockage is always explicable in terms of the ὄγκοι. Yet it is important to remember that Galen is not objecting to the admission of theoretical structures in themselves, but to their medical consequences in Asclepiades' theory.

[19] For example, respiratory problems: *De locis affectis* 8. 54-5 K; *De difficultate respirationis* 7. 781 K; hearing (and sight): *De symptomatum causis* 7. 103 K (cf. *De locis affectis* 8. 217-18 K); loss of sense of smell: *De locis affectis* 8. 214-15 K; kidney stones: *De locis affectis* 8. 18 K (and cf. 8. 403 K on inflammation). Note also the use of θρόμβωσις at Caelius, *De morbis chronicis* 4. 40.

[20] See *De morborum differentiis* 6. 856-62 K for an example of Galen's practice. Compare Oribasius' definition of στέγνωσις at *Synopsis* 5. 16. 1-4 ἡ στέγνωσις γίνεται μὲν δι' ἔμφραξιν ἢ πύκνωσιν τῶν πόρων. (Oribasius follows this with a list of signs for recognizing this blockage.)

[21] *De symptomatum causis* 7. 253 K for the idea that pores can contract through atrophy and subjection to low temperatures. As a rule, when Galen speaks of ἔμφραξις τῶν πόρων he sees the foreign bodies behind the blockage as thickened humours: see *De symptomatum causis* 7. 291-3 K. Further related cases at *De methodo medendi* 10. 473, 565-6, 602 K. The treatise *De tumoribus praeter naturam* deals generally with a wide range of such phenomena.

There is an initial question of terminology. What word did Asclepiades use to describe the particular type of impaction demanded by his theory? ἔμφραξις is the most common term for blockage in ancient medicine generally and it is not specific to any particular physiological theory. ἔνστασις is surely right, but the evidence for this reading, while certainly convincing enough for me, is not as strong as is sometimes assumed. The printed editions of Caelius Aurelianus gloss *concursus sive conventus* at *De morbis acutis* 3. 220 as ἔντασις. The ἔνστασις of Drabkin's text is a modern emendation. Similarly, Bendz has emended *in extasi* (or *extasin* in Sichart's edition) at *De morbis chronicis* 4. 6 to *in enstasi*.[22]

The manuscripts of Soranus are also divided on this point. The Asclepiadean testimonium at *Gynaecia* 3. 3. 5 reads ἔνστασις but at 3. 4. 3 the major Paris manuscript has ἔντασις, emended by Ermerins in modern editions.[23] Kühn's text of the *De methodo medendi* at 10. 101 also has ἔντασις; the ps.-Galenic *Introductio sive medicus* 14. 729 K has ἔκτασις, which the Latin translation printed by Kühn is obviously translating with *extensio*. Such mistakes can be explained easily enough palaeographically. But there could be another explanation. A Hippocratic word to describe 'tension' as a morbid condition, ἔντασις, and its related forms, σύντασις, ξύντασις, ἔκτασις, are found throughout medical literature.[24] It is conceivable that the confusion between this and the idea of ἔνστασις dates back to the original texts. The examples in Soranus and Caelius point to an even more important possibility: the similarity between ἔντασις

[22] This passage deals with Themison's treatment for elephantiasis: 'sed hoc passus est cum nondum limpide methodicam perspiceret [sc. Themison] disciplinam, et Asclepiadis secta circumsaeptus passionis causam in enstasi aestimaret quam praestabat fieri per faciem sive cutem.' The text is probably corrupt, and Bendz's suggestion (1943: 69–70) is by no means the only possible one. The emendation ultimately depends on whether 'being hemmed in by the sect of Asclepiades' necessarily means that Themison adopted an identical aetiology of elephantiasis.

[23] *Gynaecia* 3. 3. 5 [ἔνστασις] ... ταύτην γὰρ τῶν πλείστων παθῶν συνεκτικὴν (αἰτίαν) εἶναί φασιν [sc. οἱ Ἀσκληπιάδειοι]; 3. 4. 3 εἶτα καὶ τὴν ἔνστασιν οὐ πάντων, ἀλλὰ τῶν πλείστων συνεκτικὴν εἶναί φασιν.

[24] This is perhaps what Hicks (1962), 390, had in mind when he translated ἔνστασις νοητῶν ὄγκων ἐν νοητοῖς ἀραιώμασιν at Sextus Empiricus, *Adversus mathematicos* 8. 219–20, as 'a tension of corpuscles in interspaces'.

[ἐντείνω] and *strictura* [*stringo*] is obvious, and we may have here some kind of subconscious assimilation of Asclepiades to Methodism by some of the Methodists themselves. I think it more likely that the concept of ἔνστασις itself developed into one of the so-called Methodist 'communities' (more of this later). ἔνστασις is, I suggest, a theory-specific term, and it is possible that it was used in a medical sense for the first time by Asclepiades.[25]

There is little evidence from sources who are either not Methodists, or who do not, like Galen, have a vested interest in insisting that Methodism is little more than watered down Asclepiadeanism. Where such evidence does exist, it is unambiguously behind ἔνστασις. A rather obscure figure of indeterminate date, Cassius the Iatrosophist, consistently uses this term where Asclepiades' theory, or one very similar to it, is in play.[26] Most telling, perhaps, is the way in which Caelius consistently renders the word with Latin equivalents such as *obtrusio*, *statio* (and *coacervatio*).[27]

Pathogenic ἔνστασις, then, is an interruption of flow of the ὄγκοι through the πόροι. Celsus, never particularly interested in the details of the theory and never particularly involved in medical squabbles, puts it succinctly: illness comes about when 'corpuscles which flow through invisible pores close off their path by piling up'.[28] Regardless of where it takes place, and

---

[25] It should be said, however, that the term seems to have found its way into the excerpts from Herodotus the Doctor made by Oribasius, where it is used apparently without the Asclepiadean overtones. (See Oribasius, *Collectiones Medicae* 5. 30. 5.)

[26] Cassius, *Problemata* 76 ἔνστασις δέ ἐστιν ὄγκος ἐν λόγῳ θεωρητοῖς ἀραιώμασι διὰ σφήνωσιν; see also *Problemata* 77, 79.

[27] These Latin words are glossed with *emphraxis* by Cassius Felix: *De medicina* 44, p. 110. 4 'ad conclusionem sive obstrusionem [*vel* obtusionem] epatis, quam enfraxin vocant.' Also Caelius, *De morbis acutis* 3. 64, on *conductio*: 'alii nostrae sectae scriptores pro tensione conclusionem vocaverunt, sub alio nomine id ipsum significantes'; Cassius, *De medicina* 28, p. 46. 1 'ad tinnitus et obstrusiones aurium, quos Graecis enfraxis vocant'. Bendz (1964), 24 ff., suggests that the double expressions we find in Caelius, such as *concursus sive conventus*, *statio sive obtrusio*, and so on, are all translations of the one Greek term. See Bendz (1964), 86, for a discussion of the orthography of *obstrusio/obtrusio/obtusio*.

[28] *De medicina* 1, proem 15-16 'manantia corpuscula per invisibilia foramina subsistendo iter claudunt'. Note Celsus' use of *mano* to describe the way in which the corpuscles move; it is a word generally applied to the movement of fluids.

regardless of the effects it produces, impaction is always of one kind: of theoretical corpuscles in theoretical pores. The impaction itself may come about through the obstruction of passages by a build-up of corpuscles passing through, as Celsus suggests, or through some alteration in the state of the pores themselves.[29] Let us begin with the first case. Sextus Empiricus, *Adversus mathematicos* 3. 5, provides us with a convenient starting-point for understanding the mechanics of this type of blockage.[30] Building on the interpretation which I developed in Chapter 1, we may regard it as a list of theoretical conditions which need to be met if blockage is to occur.

The first condition, that there must be pores of differing sizes in our bodies, explains how corpuscles can move about inside us. The fact that the pores come in a variety of different shapes and sizes is important; impaction is more likely to occur in parts of the body where the pores are fine, or not straight. The second condition explains the origin and nature of the fluids which come to be concentrated in the affected parts. Sextus' phrase, 'the parts of moisture and pneuma' (i.e. the corpuscles), is almost certainly equivalent to the terms Caelius applies to Asclepiadean fluids, such as *suci*,[31] *liquidus atque spiritus*,[32] and to the 'vapours' which pass into the bladder in Galen's account of the Asclepiadean theory.[33] When their passage is blocked, impaction results. The third and final of Sextus' conditions is rather differ-

---

[29] This latter course is outlined at Caelius, *De morbis acutis* 1. 107, in a difficult passage which I shall discuss shortly.

[30] The text is quoted above, p. 26.

[31] *De morbis acutis* 1. 106.

[32] Ibid. 107.

[33] Galen, *De naturalibus facultatibus* 2. 30–44 K. Caelius' description of the passages of corpuscles through the pores at *De morbis acutis* 1. 106 ('sucorum ductus solito meatu percurrens') is certainly equivalent to Celsus' 'manantia corpuscula per invisibilia foramina'. There is almost certainly some trace of a humoral pathology here. I have already noted Celsus' use of *mano*, which implicitly points to a blockage by *fluids* in the body. Yet Caelius hints at an Asclepiadean rejection of the argument that the humours directly cause disease at *De morbis acutis* 1. 112: 'et non esse activas atque operantes causas aegritudinum in liquidis constitutas, quas synecticas vocant, sed esse antecedentes, quas Graeci procatarcticas appellant.' What seems most likely is that Asclepiades preserved some of the traditional language of humoral pathologies, while totally abandoning the causal analyses of disease which came with them. On the variety of humoral pathologies on offer, see Schöner (1964).

ent: 'There are continuous emanations from our bodies to the outside.' Wellmann believed that Sextus was hinting here at a doctrine of skin-breathing, and he may well have been right.[34] More importantly, I suspect that there is a reference here to the mysterious flux theory which is mentioned by a few witnesses, including Cassius the Iatrosophist and two later Galenic commentators. A number of reports state that Asclepiades held the whole world, like its component parts, to be in a constant state of motion. It seems that a variety of epistemological and methodological positions were derived from this, and notably, from the notion that 'nothing can be seen twice in the same way'.[35] The flux theory turns out to be a corner-stone of Asclepiades' pathology.

It will come as no surprise by now to learn that the evidence is thin. Sextus Empiricus mentions it rather vaguely at *Adversus mathematicos* 8. 6–7, along with a more familiar Platonic argument that because the perceptible world is always coming to be and never being, only τὰ νοητά are true:

οἱ δὲ περὶ τὸν Πλάτωνα καὶ Δημόκριτον μόνα τὰ νοητὰ ὑπενόησαν ἀληθῆ εἶναι, ἀλλ᾽ ὁ μὲν Δημόκριτος διὰ τὸ μηδὲν ὑποκεῖσθαι φύσει αἰσθητόν, τῶν τὰ πάντα συγκρινουσῶν ἀτόμων πάσης αἰσθητῆς ποιότητος ἔρημον ἐχουσῶν φύσιν, ὁ δὲ Πλάτων διὰ τὸ γίγνεσθαι μὲν ἀεὶ τὰ αἰσθητά, μηδέποτε δὲ εἶναι, ποταμοῦ δίκην ῥεούσης τῆς οὐσίας, ὥστε ταὐτὸ μὴ δύο τοὺς ἐλαχίστους χρόνους ὑπομένειν, μηδὲ ἐπιδέχεσθαι, καθάπερ ἔλεγε καὶ ὁ Ἀσκληπιάδης, δύο δείξεις διὰ τὴν ὀξύτητα τῆς ῥοῆς.

The followers of Plato and Democritus suppose that only the intelligibles are true; yet Democritus says that this is because nothing is naturally perceptible, and the atoms which combine to form the universe are themselves devoid of any perceptible quality. Plato, on the other hand, says it is because the perceptible world is always coming-to-be, and never being—it is like the water in a flowing river, not remaining the same for two instants; as Asclepiades also said, it cannot be pointed out twice on account of the speed of the stream.

[34] Wellmann (1901), 82 ff.
[35] For the dictum that 'nothing can be seen twice in the same way', see Galen, *De sectis* 1. 75–6 K; it is discussed at great length in the first seven chapters of *On Medical Experience*. This doctrine has nothing directly to do with the Methodist community of *flux*.

What is behind Asclepiades' assertion that you cannot point
at the same river twice because of the speed of its flow? Heraclitus need not be summoned directly, although he is probably
somewhere in the distant background.[36] The juxtaposition of
Plato and Asclepiades might be thought a more useful clue;
flowing rivers feature in several physiological analogies in the
Timaeus.[37] But in my view such similarities with earlier philosophical theories only cloud the more immediate background to
the Asclepiadean doctrine.[38] Two ostensibly less reliable sources
may hold the key.

First, then, Cassius the Iatrosophist. In the first of his *Problemata* he poses (and, of course, answers) a rather arcane question: 'Why are round wounds harder to heal than others?' Some
Herophileans, he says, give one solution: the larger the surface
area of a wound, the longer it takes for cicatrization to complete.
Round wounds are often larger than they look, as simple
geometry will show. Cassius then brings in the rather different
account of Asclepiades. I translate the relevant part below:

αὐτὸς δὲ ὁ Ἀσκληπιάδης τοιαύτην ἀποδίδωσιν αἰτίαν· φησὶ γάρ, ὅτι
ἐπὶ παντὸς πράγματος, ᾧ συμφυτόν ἐστι τὸ κινεῖσθαι, ἡ σφοδροτέρα
κίνησις γίνεται ἐκ τῶν κατ' αὐτὰ ἀρχῶν. ἀρχὰς δὲ λέγει τὰ μέσα τῶν
κινουμένων. ὑποδείγμασι δὲ χρῆται πρὸς σαφήνειαν τῇ τε τῶν
ποταμῶν κινήσει καὶ τῇ τοῦ πυρός. ὡς γὰρ ἐπὶ τούτων μάλιστα
κινεῖται καὶ σφοδρότερον τὰ μέσα, οὕτω συμβαίνει καὶ ἐπὶ τῶν ἑλκῶν·
οἷόν τι ἐστι τὸ λεγόμενον, ἐπεὶ τῶν ποταμῶν ἐπὶ τοῦ πλεονάζοντος
μέρους σφοδροτέρα γίνεται ἡ κίνησις. πλεονάζει δὲ δηλονότι τὸ μέσον
ἐν αὐτοῖς· ἐκεῖ γὰρ μάλιστα συντρέχει τὸ πλέον. ὁ δ' ὅμοιος λόγος καὶ
περὶ τῆς φλογὸς τοῦ πυρός. κατὰ γὰρ τὸν μέσον μάλιστα τόπον
σφοδρότερον κινεῖται αὕτη οὕτως, ὡς καὶ πολλάκις προαναπηδᾶν
κατὰ τοῦτο τὸ μέρος τὴν φλόγα. ἐπεὶ οὖν καὶ τὰ στρογγύλα ἕλκη,
καθότι αὐτὰ ὥσπερ συνῆκται, καὶ οἷον εἰς μέσα ἐστὶ κατὰ πᾶν μέρος,

---

[36] Although [Aristotle], *Problemata* 908ᵃ28–ᵇ10, associates 'some Heracliteans' with a
medical problem. The Epicurean Diogenes of Oenoanda attributes to Aristotle a flux
theory where the speed of the flux seems to be important, at pp. 8–9 Chilton. Flux
theories can function in sceptical systems just as in rational systems.

[37] e.g. 43 B–C, 77 C, 79 A, 80 D.

[38] It is only fair to mention Stephen of Athens' commentary on Hippocrates' *On
Prognosis*, p. 67 Dietz (1834): καλῶς γὰρ ὁ Πλάτων ποταμῷ ῥέοντι ἀπείκασε τὰ ἡμέτερα
σώματα. καὶ διὰ τοῦτο οὐκ ἔχει τὸ βέβαιον ἡ τέχνη.

διὰ τοῦτο συμβαίνει σφοδροτέραν τὴν κίνησιν ἐπ' αὐτῶν γίνεσθαι. κίνησις δὲ γίνεται δηλονότι ὄγκων τινων φυσικῶς φερομένων ἐπὶ τὰ ἔξω διὰ τῶν πόρων, καὶ τῇ σφοδρᾷ τούτων παρόδῳ συμβαίνει ἀνακρούεσθαι τὴν ἐπούλωσιν. ῥητέον δὲ πρὸς τοῦτο, ὅτι εἰ τῶν ὄγκων ἡ σφοδρὰ κίνησις ἐπὶ τὸ ἔξω αἰτία γίγνεται τοῦ εἶναι τὰ τοιαῦτα ἕλκη δυσαλθῆ, ἐχρῆν μάλιστα τοῦτο συμβαίνειν ἐπὶ τῶν ἀκμαζόντων· καὶ ἐκ τούτων ἂν συμβαίη ἕλκη ἐπ' αὐτῶν τὰ τοιαῦτα· ἐπὶ γὰρ τούτων μάλιστά εἰσιν ὀξυκίνητοι οἱ ὄγκοι καὶ πολλοὶ τῷ πλήθει, ἐπὶ δὲ τῶν γεγηρακότων ὀλίγοι τέ εἰσι καὶ νωχελεῖ τῇ κινήσει χρῶνται.

Asclepiades gives the following kind of explanation: he says that in general, where bodies possess innate movement, the more vigorous movement takes place at the origins, and that the origins lie at the centre of bodies which are moved. To clarify this point, he uses the example of rivers and of fire. For in these cases, the middle parts are moved more vigorously; this is also the case with wounds. What this means is that in the case of rivers, more vigorous movement takes place in the part of the river which contains most—in the middle. A similar argument applies to the flame of fire. For there is vigorous motion especially at the centre; consequently the flame frequently leaps up at this point. And so it is with round wounds: they are similarly constituted and are in every part related to their central points, and thus it happens that with them too there is more vigorous motion (at their centre). The motion has its origins in the natural conveyance of corpuscles out of the body through the pores; the force of their exit inhibits cicatrization. We must add to this, that if the vigorous passage (of corpuscles) out of the body is the cause of such wounds being slow to heal, then we should expect to find that this is especially the case with patients who are in their prime. And indeed such wounds will tend to occur in this age-group. For in them the corpuscles are particularly rapid in their movement, and there are large numbers of them, while in old people they are few and sluggish in their motion.

The river in this analogy is meant to illustrate the flow of corpuscles through the pores. The natural efflux of corpuscles from the body corresponds to the flow in the third of Sextus' hypotheses. A little more information, which brings the reports of Cassius and Sextus closer together, comes from some very unlikely sources. John of Alexandria and Agnellus of Ravenna, in their much later commentaries on Galen's *De sectis ad introducendos*, give some idea of the theory's dialectical reception in the debate between Asclepiades and the medical empiricists: I pick up the commentary on *De sectis* 1. 75–6 K:

Dogmatici prius accusant empiricos tribus modis quia inutiles sunt et indocti et inconstantes. etenim Asclepiades inter eos primus dixerat quia nihil in semet ipso manet, sed omnia immutantur et corpora nostra flumini assimilantur, quod sine intermissione fluit et semper non unam poteris tangere aquam quia quam modo tetigisti fluens perlabitur unda, et iterum quia infantum alba et molliora sunt corpora, at ut creverint nigriora fiunt, et quando barbam miserint, facies immutatur.

Before I translate this passage, I should explain myself a little. Both commentators seem confused about Asclepiades, and suggest, at first sight, that he was an empiricist, but the context in Galen makes it quite clear that he is one of the dogmatists attacking empiricist medicine. Galen has found three rationalist physicians opposed to medical empiricism who assail it on three different grounds: it is ἀσύστατος (incoherent), ἀτελής (inchoate) and ἄτεχνος (artless). Galen has Asclepiades present reasons for the first attack, Erasistratus for the second, and Athenaeus for the third. The commentator seems to be rendering ἀσύστατος with *inconstans*, ἄτεχνος with *indoctus*, and ἀτελής with *inutilis*. He elaborates on the Asclepiadean argument, introducing ideas hinted at by Sextus Empiricus but not present in the Galenic account. In spite of appearances, then, I believe we have evidence here that Asclepiades attacked the whole idea that experience in medicine can be useful, on the ground that the world is in such a state of flux that we cannot be sure that yesterday's experience will be relevant tomorrow.

The rationalists attack the empiricists in three ways—because they are useless, ignorant, and inconsistent. First among them, Asclepiades said that this is because nothing remains in its own place, but everything changes, and our bodies are like a river which constantly flows. You can never touch the same water twice, because the part you touch has flowed on. Similarly, the bodies of children are white and soft, but as they grow, they grow darker. When they grow beards, their very appearance changes.[39]

At least one significant point arises from the last few lines: it seems that Asclepiades gave the example of changes in the human constitution as one effect of the underlying flux in the body. The example given recalls those provided by Caelius Aurelianus at De morbis acutis 1. 106; examples which are generally taken to be redolent of Epicurean atomism. Here, Caelius reports an Asclepiadean argument that the part does not have to resemble the whole—that horn shavings are not the same colour as horn, nor silver filings as silver. These examples may well be atomistic in origin, but we must accept the possibility at least that they were deployed by Asclepiades in a quite different way. If the stress placed on movement of the corpuscles, and consequent impermanence of the different visible agglomerations which their movement generates, *does* involve the flux theory, then this might explain one problem of apparent affiliation with atomism which I left unresolved in Chapter 2.

These passages illuminate a number of other troublesome matters, most notably the Asclepiadean doctrine on the substance of the soul. There is ancient agreement that the soul for Asclepiades was 'generated' through respiration. He saw in it no 'ruling faculty' (ἡγεμονικόν in the philosophical jargon of the time), but regarded it in purely physical terms.[40] Now we can see why. The substance of the soul for Asclepiades was pneuma—very fine corpuscles which permeate the living body and which originate in inspired air.[41] The fact that interference with

---

[39] John of Alexandria, *Commentarium in de sectis Galeni* 4va3-10 Pritchet. Agnellus of Ravenna (*Commentarium in de sectis Galeni*, p. 78. 10-21) has the same account, but reads *similant corpora nostra* ... for *assimilantur* ... Yet it is clear that the analogy is Asclepiades', and not the empiricists'.

[40] Galen, *De usu respirationis* 4. 471 K, notes that according to Asclepiades we breathe for the sake of the generation of the soul. At *Commentarium VI in librum VI Hippocratis Epidemiarum* 17B. 319-20 K he states more specifically that Asclepiades believed that the psychic pneuma was generated by breathing, while for Erasistratus it was the vital pneuma. There is no evidence that Asclepiades distinguished between vital and psychic pneuma; it is very likely that he modified the Erasistratean doctrine on this matter.

[41] On the Asclepiadean denial of the existence of the ἡγεμονικόν see Sextus Empiricus, *Adversus mathematicos* 7. 380. On the nature of the constituents of the soul,

breathing causes death neatly underlines this connection. It seems reasonable to suppose that the flux of fine corpuscles which, according to Cassius, continually leaves the body has its origin in inspired air; the more vital and youthful the body, the stronger the flux, and therefore the stronger the soul and ultimately the whole constitution. This is not to say that respiration is the cause of the flux; merely an example of it at work. Strictly speaking, Asclepiades' whole system requires that everything is in constant flux, and this flux can be related to the natural motion of the corpuscles. Certain aspects of this theory may well owe something to the account of respiration through the skin in the *Timaeus*; its distant debt to some kind of Heraclitean theory is obvious. But Erasistratus is also a possible source for the idea that the flux involves invisible effluences.

Against this background, we may now investigate how the concept of ἔνστασις was actually used to explain disease. The best-documented case is that of phrenitis; it occupies much of the first book of Caelius' *De morbis acutis*, and an obscure work *De morbis acutis et chronicis*, identified by Fuchs as being by Themison; Asclepiades is not mentioned in this last treatise.[42] Phrenitis itself has not been identified positively; even in antiquity there was disagreement over the affection it signified. That does not really matter. Caelius, like Asclepiades, understands it as an affection based in the *mentes*. At *De morbis acutis* 1. 6 he gives what he implies is a quotation of the Asclepiadean definition: '*phrenitis* ... is an impaction or blockage of corpuscles in the cerebral membranes which occurs frequently without any sympathetic affection, but with mental alienation and fever.'[43] Again at 1. 20 he quotes from Asclepiades' work *On Definitions*: 'He says that phrenitis is a sudden alienation accompanied by fever.'[44]

Calcidius, *In Platonis Timaeum* 215, p. 229 Waszink: 'aut enim moles quaedam sunt leves et globosae ... ex quibus anima subsistit, quod totum spiritus est, ut Asclepiades putat.' The substance of the soul is, of course, cognate with τὸ λεπτομερές of the Greek reports.

[42] Fuchs (1903), 69–70.

[43] 'phrenitis ... est corpusculorum statio sive obtrusio in cerebri membranis frequenter sine consensu, cum alienatione et febribus.'

[44] 'phrenitim inquit esse alienationem repentinam cum febribus.'

Caelius is not particularly interested in the *statio corpusculorum* and its nature. He directs some rather sceptical arguments against it, first casting doubt on whether those wretched corpuscles exist or not, then highlighting disagreements between followers of Asclepiades over what their master meant. His immediate criticisms centre on more general issues of ambiguity in Asclepiades' expression and his omission of (what Caelius believes are) important physical symptoms. The overall criticism is quite straightforward: Asclepiades wasted his time attempting all these explanations. He privileged the causes over the effects of disease, and for the Methodist physician the visible signs presented by a disorder are all the doctor needs to worry about.[45]

Galen is rather more interested in attacking the theory behind these explanations; in the mouth of the empiricist in the treatise *On Medical Experience* (28. 3) he gives what appears to be a quite different account of Asclepiades' *phrenitis*. He mentions none of the symptoms described in Caelius' text, but does discuss the development of the pathogenic blockage in some detail:

For you say: 'burning fever inflames the cerebral membrane, and it results from this that the atoms make their way to the finely divided thing, or those of them which do so become extremely fast and violent in motion all at once; this is followed by a stoppage of the atoms in the pores which causes the disease known as phrenitis. Thereupon what lies beneath the cartilages spreads upwards, being attracted by the finely divided thing (τὸ λεπτομερές). Now when the very numerous atoms rise and rub against the resisting parts, they are repelled. After this, they return to the roomy parts which are capable of absorbing

---

[45] Caelius, *De morbis acutis* 1. 7–8. Caelius' own 'understanding' of the disease (he 'understands' things, and does not define them—a typically Methodist trait) shows just where the disagreement with Asclepiades lies: *De morbis acutis* 1. 21 'phrenitim esse alienationem mentis celerem cum febri acuta atque manuum vano errore, ut aliquid suis digitis attrectare videantur, quod Graeci crocydismon sive carphologian vocant, et parvo pulsu ac denso' ('Phrenitis is an acute alienation of the mind, accompanied by fever and random movement of the hand which results in a clawing motion. The Greeks call this *crocydismon* or *carphologia*. The pulse is small and heavy'). Compare Caelius' attack on the Asclepiadean explanation of diarrhoea at *De morbis acutis* 3. 221, Leonides the Epi-synthetic's definition of lethargy at 2. 7–8, and Apollonius Mys' definition of pleuritis at 2. 88–9.

them, and for this reason the belly is loosened. Since this is the case, it is therefore necessary for the origin of the burning fever and its accompanying symptoms to come first, after which phrenitis follows, then comes the upward attraction of the regions of the cartilages and the phrenitis is followed by the loosening of the belly.' (trans. Walzer.)

We might summarize the main points in the process thus:

1. The cerebral membranes are heated.

2. Corpuscles begin to move towards the cerebral membranes because of the heat there.

3. Crowding of the corpuscles ensues, and this causes a blockage in the cerebral membranes, which Asclepiades calls *phrenitis*.

4. The movement towards the cerebral membranes continues (PTLP in action), and before long they can accommodate no more. Some of the corpuscles are repelled, and they then make their way to the parts of the body which can accommodate them. This results in the condition which is called 'loosening of the belly'.

We can build up a fuller picture of the pathology around this outline.

1. *The cerebral membranes are heated*

Walzer translates the Arabic here: 'burning fever inflames the cerebral membrane.' It is important to remember that concepts of 'fever' varied from doctor to doctor. There was, for example, considerable disagreement over whether or not fevers should be counted as diseases in their own right rather than straightforward 'symptoms' of other affections.[46] When he is talking about Asclepiades, Caelius uses the term *febris* to cover a variety of manifestations of heat in the body, ranging from serious fevers brought about by different types of impactions in the pores, to milder fevers with their origin in large-scale 'disturbances'. And so Caelius can say: 'quomodo febrem illam vehementam dixit

---

[46] Diocles, for example, held fever to be a sign of disease: Aetius, *Placita* 5. 29. 2.

[sc. Asclepiades] ... quae per stationem vel obtrusionem fuerit, illam vero solubilem atque levem quae ex turbatione fuerit liquidarum materiarum atque spiritus' ('Asclepiades says that a violent fever is one caused by an impaction or obstruction, and that a "loose" (*solubilis*) and light fever originates in a disturbance of liquid matter and pneuma').[47] As a general rule, Asclepiades seems to have explained all bodily heat in terms of the friction of moving corpuscles against each other and against surrounding structures. When Galen claims that Asclepiades reduced all fevers to ἐμφράξεις[48] he is probably just thinking of the more serious types.

The initial heating of the cerebral membranes which ultimately leads to phrenitis, though, cannot have been brought about by a blockage—induced fever—for reasons I shall outline under the next heading. Caelius maintains that Asclepiades' phrenitics suffered from fever, but the language of the definition at *De morbis acutis* 1. 6 implies that this fever follows on from the *statio*. He is still more specific about this at *De morbis acutis* 1. 123-4: phrenitics suffer from fever, and this fever is caused by a stoppage of the larger corpuscles. At *De morbis chronicis* 1. 146 he uses the presence of fever to distinguish phrenitis from other disorders which affect the mind, and which are thus centred on the cerebral membranes.[49] The fever in phrenitis to which Caelius refers, then, will be the result of the impaction. Some distinction needs to be made between this and the initial heating of the membranes.

A 'high temperature' can be a sign of fever, but the body can be heated in other ways; it comes as no surprise that some of the

---

[47] *De morbis acutis* 1. 8. Caelius *may* be pointing to confusion over this matter when he attacks Asclepiades for using the corpuscular hypothesis to explain 'fever' when what he is really explaining is the state of 'being feverish' at *De morbis acutis* 1. 11.

[48] Galen, *De tremore, rigore et palpitatione* 7. 615 K; see also *De diebus decretoriis* 9. 798 K; Asclepiades is not mentioned here—in a rare pun, Galen introduces the 'Asclepiads'—but is certainly behind Galen's attack on those people who dream up 'theoretical' fevers which are the result of 'theoretical' particles.

[49] See also *De morbis acutis* 1. 15: the fever is an integral part of phrenitis. In cases where there is *alienatio* without fever, the result is called *mania* or *furor*. And *De morbis acutis* 3. 112, where Caelius notes that 'secundum Asclepiadem omnis passio quae mentem turbaverit in ipsa consistat, ut phrenitis, lethargia, epilepsia'.

other ways feature in the beginning of the Asclepiadean aetiology. Hot weather can predispose people to phrenitis, as can the
kind of mental disturbance which involves the dilation of pores
in the brain, the rapid movement of the corpuscles there, and
consequent heating by friction. Factors like these were put
under the general heading of antecedent causes by many non-
Methodist doctors. Asclepiades, for one, recognized their
importance.[50] Even in the Asclepiadean discussion of dislocation
at Oribasius, *Collectiones medicae* 47. 13, he remarks that the cases
occurred in Parium. All this is anathema to Caelius: at *De morbis
acutis* 2. 130 he says 'apud Romam . . . utemur phlebotomo, nulla
regionum discretione confusi' ('we shall employ venesection at
Rome, without any distinction based on locality'). Even Galen
confirms this aspect of Methodist practice; in the *De sectis* 1. 79–
83 K he sees it as a serious flaw. He indulges in an attack on the
Methodists for refusing to accommodate antecedent causes at
Adversus Iulianum 18A. 255 K. Here he even praises Asclepiades
for his admission of the importance of such types of information—something Galen hardly does often. Elsewhere, Caelius
adds indigestion, intoxication, exercise taken after meals, and
time spent in places with a 'heavy' atmosphere to the list of
Asclepiadean signs pointing to the likelihood of phrenitis.[51]
Signs of a likely lapse into disease could also be derived from
factors related to the constitution of the patient, such as age.[52]
Taken individually, these do not necessitate the disease;
together they suggest a strong probability.

---

[50] *De morbis acutis* 1. 14–15; for the effects of hot weather and other environmental
factors, see 1. 32. At 1. 139 Caelius notes that Asclepiades 'pro temporum aere atque loco
ordinanda adiutoria putat'. 2. 63–4 contains Asclepiades' observations of particular types
of quotidian fevers he has seen at Rome; 2. 129 finds him taking the locality into account
when deciding whether or not to phlebotomize pleuritics. (And phrenitics, at *On Medical
Experience* 26. 5–6.)

[51] *De morbis acutis* 1. 22, 1. 141. Caelius himself insists that the only signs pointing to a
disease can be those which necessarily do so in all cases. See *De morbis acutis* 1. 22–3, and
2. 176 for the view of Soranus: 'nam signum neque recedit, et semper significato coni-
unctum est; accidens autem, quod Graeci σύμπτωμα vocant, nunc advenit, nunc recedit.'

[52] The flux in young people is faster than in the old, and this is related by Cassius the
Iatrosophist to the intensity of an illness at *Problemata* 1.

2. *Corpuscles begin to move towards the cerebral membranes because of the heat there*

PTLP is responsible for this movement; here we have an example of a ῥύσις ἐπὶ τὸ θερμαινόμενον. Bearing in mind the *fervor/λεπτομερές* connection made by Caelius, we can imagine an area of λεπτομέρεια being formed in the cerebral membranes as a result of the heat there—the heat seems to cause an area of rarefaction to develop.[53] The state of λεπτομέρεια puts into action the chain of events which induces blockage; this is why it is unlikely that Asclepiades could have envisaged the initial heating as itself the result of a major blockage in the membranes around the brain. Anything which encourages or stimulates this state should be avoided in treatment. And so we find, for instance, that cutting the hair of the phrenitic patient is condemned. This would merely encourage the flow of matter to the head, and consequently make the blockage still worse.[54]

3. *Crowding of the corpuscles causes blockage*

The blockage in the cerebral membranes is the central feature of Asclepiades' characterization of the disease. (Note that Caelius speaks of a 'blockage in the cerebral membranes', while Galen is more theory-specific, and talks of a blockage 'in the pores of the cerebral membranes'.) At this level, phrenitis is only distinguishable from other diseases involving blockage in terms of the location of the problem. But to judge from the pitiful evidence available for other diseases, this seems to be a typical explanation of impaction. A change in state somewhere in the body brings about a PTLP—induced flow of matter towards the part which

---

[53] These membranes are noted by Galen for their spongy nature at *De usu partium* 3. 651-2 K; they are πολύτρητα and σηραγγώδη, which would make them roughly similar in texture to the lungs (as far as the Asclepiadean theory is concerned).

[54] *De morbis acutis* 1. 116 'incusat ... tonsuram, siquidem hortamento quodam liquidorum faciat ad caput ascensum, et in constrictionem atque tensionem cerebri membranam cogat'. The same idea, without its author, reappears at ps.-Alexander, *Problemata* 2. 36 διὰ τί ἐπὶ ὀφθαλμίᾳ ξύρησις ὠφέλιμος; ὅτι διαπνεομένη ἡ κεφαλὴ διὰ τῶν πόρων διαφορεῖ τὰ περιττώματα καὶ τὸ ῥεῦμα.

was initially affected. The movement continues until the affected part can accommodate no more.[55]

Blockage by corpuscles of different sizes gives rise to different effects, as does their number, shape, and speed in the affected area. Severe fevers are brought about by the impaction of the larger corpuscles in the larger pores. We might imagine that the fevers mentioned by Caelius as concomitant features of phrenitis are the result of larger particles moving towards the head. The movement of smaller particles, excited by whatever antecedent cause is in play, could then be used to explain the initial heating of the membranes.

I have already suggested that Asclepiades classified the mobile fluids in the body under the general heading 'liquids' and 'pneuma'. Caelius tells us that the 'larger' corpuscles made up blood, and the smaller, finer ones made up pneuma.[56] What in effect seems to be happening is a migration of fluids to parts of the body where they are not required. This is a doctrine which I shall presently trace back to Erasistratus. It is the movement which is, in one sense, the real centre of the Asclepiadean explanation of disease.

Phrenitis, then, provides us with one example of how an impaction can occur. Caelius has a few more general details about the nature of blockage in the pores in his main account of Asclepiades' theory (*De morbis acutis* 1. 107): the text is problematic, and I shall explain my reasons for the translation which follows presently:

fit autem eorum [sc. corpusculorum] statio aut magnitudinis aut schematis aut multitudinis aut celerrimi motus causa, aut viarum flexu †conclusione† atque squamularum †exputo† varias inquit [sc. Asclepiades] fieri passiones locorum aut viarum differentia.

---

[55] Similarly, Cassius the Iatrosophist speaks thus of the blockage which causes headache at *Problemata* 77: ἡ τῆς ὕλης φορὰ γίνεται καὶ τὰς ἐνστάσεις αὔξει. Asclepiades is not mentioned here, but Cassius is clearly expressing Asclepiadean ideas in Asclepiadean language.

[56] See *De morbis acutis* 1. 124: a discussion of why, according to Asclepiades, venesection is lethal for phrenitic patients: it withdraws both the larger and the smaller corpuscles, that is to say not only blood but pneuma and heat too. Cf. *De morbis chronicis* 3. 65.

The first part of this sentence presents no problems. We have already seen how different-sized corpuscles can underlie and affect different functions, and how the speed of their motion can influence the gravity of an attack. But the phrase 'aut viarum flexu conclusione atque squamularum exputo' has been universally condemmned by editors of the printed editions. (Unfortunately there are no surviving manuscripts of this passage.) It is far from clear to me how *conclusione* fits in syntactically, and it may in fact be no more than an intrusive marginal gloss on 'viarum flexu'. Or else we might read something like 'viarum flexu sive conclusione', understanding *conclusio* as a property of the pores, and *statio* of the corpuscles. Andernach in his edition of 1533 noted a lacuna between *squamularum* and *exputo*; Almeloveen, in Amman's edition of 1709, noted a hiatus too, and made an extraordinary suggestion to fill it: *squamularum obtrusione, ut in retento sputo*. Drabkin obelized the phrase and relegated his own suggestion, *conclusione corpusculorum effecto*, to the apparatus. At any rate, the *squamulae* are most probably equivalent to the *fragmenta*, and even if I cannot render the Latin exactly, this is what I think it means.[57] My translation runs thus:

An impaction of the corpuscles comes about as a result of their size, shape, number, or speed. Alternatively it can occur as a result of bending of the pores [. . .], or [shedding?] of fragments (lit. 'scales').

If I am right in saying that these 'scales' are to be identified with the fragments (and I am by no means the first to have done this), then they may represent the result of the disintegration of the walls of the passages—the *solutio* of the corpuscles which make up the pores.

There are some scattered references to ἐνστάσεις brought about by contraction of the pores themselves, although there is no evidence beyond the passage above for how the process might have come about. At one point Cassius the Iatrosophist treats an 'occlusion of the pores' as equivalent to ἔνστασις,[58] and

---

[57] David Sedley has suggested that some form of a verb *exspuo* might be lurking here.

[58] *Problemata* 79 διὰ τί οἱ μαινόμενοι παρακόπτουσιν; ὅτι δυσοδία τοῦ ψυχικοῦ πνεύματος ἀποτελεῖται διὰ τὴν ἔνστασιν καὶ τὴν μύσιν τῶν ἐν τοῖς μεταβάλλουσιν αὐτὸ

the idea of passages constricting (as opposed to being blocked by peccant bodies) is common even with other doctors who did not share Asclepiades' theory—Galen, for instance, as we have already seen (though this is against the background of his own pathology): pores may be blocked by humours passing through them, or may block up when they atrophy, are cooled, or are softened. As I have said, Galen's language here is often remarkably close to that of Asclepiades. Galen has many ways of distancing himself from theories which might be thought superficially similar to his own, but he is generally aware of the fine line he is treading. In a passage in the *De methodo medendi*, in a context where there is no mention of Asclepiades (but which is aimed at the Methodists), he argues that those who talk in terms of balance and lack of balance in the pores are using an inferior way of talking about density and rarefication.[59]

4. The movement continues... and results in the loosening of the belly

This is in many ways the most difficult stage of the process to grasp. It is also one of the most important, for here lies the true flexibility and sophistication of Asclepiades' theory. Here also lies the key to understanding how his successors took and developed the corpuscular hypothesis. Galen's account of phrenitis states that the movement of corpuscles towards the cerebral membranes continues until they can take no more. When this happens, the corpuscles move back to parts of the body which have the room to accept them. There they give rise to another, different affection—an ominous-sounding 'dissolution in the belly'. This is a basic principle which I have already described in action in the theory of respiration. In that case, the lungs are filled to capacity with inhaled air before they reject what they cannot accommodate. The process is still more clearly outlined in Cassius the Iatrosophist's explanation of how secondary affec-

σώμασι πόρων. Cf. *Problemata* 82 διὰ τί οἱ δυσπνοοῦντες ἦχον συριγματώδη ἠχοῦσιν; ὅτι μύσις τις ἐστι καὶ σύμπτωσις πόρων τῶν ἐν τῷ πνεύμονι πάθος.

[59] Both terms, he believes, are specific to continuum theory: *De methodo medendi* 10. 770-1 K.

tions can occur in parts of the body quite remote from the centre of the primary affection.[60] (In this last case the affection is an injury rather than a disease.)

Asclepiades seems to have made some use of the concept of remote, yet sympathetic, affections. It enabled him to explain even the more intractable phenomena of disease. Take diarrhoea, for instance. Caelius claimed to be nonplussed at the Asclepiadean view that diarrhoea is the result of a blockage of corpuscles.[61] Surely diarrhoea is an affection characterized by anything but blockage? Caelius does not specify exactly where Asclepiades thought this blockage takes place, but it does follow that the corpuscles should gather in the 'lowest parts of the colon', whence the flux itself actually comes. But in the light of the process reconstructed above, it would be reasonable to imagine that a blockage gives rise to a sympathetic affection whose characteristic is the flow from the bowels called diarrhoea. Perhaps Asclepiades found that many phrenitic patients were afflicted in this way.

THE 'SOLUBILES'

Now the territory becomes still less clear. There are a few cases where a dissolution or flux in the body was not a sympathetic affection, a *consensum* (to use what I think is the Caelian term), of a blockage-induced affection. These cases can be relegated to the group of diseases which, according to Caelius and Soranus, were not attributed to blockage at all. They do not form a par- ticularly coherent group, and there is no detailed evidence describing them; much of what follows is inevitably rather con- jectural. What does seem certain, however, is that diseases in this class owed their origin to 'disturbance' in the body. Caelius again provides the most detailed account:

et non omnes statione corpusculorum sed certas, hoc est phrenitim, lethargiam, pleuritim et febres vehementes; solubiles vero liquidorum

---

[60] *Problemata* 40. See above, pp. 86–7.   [61] *De morbis acutis* 3. 220.

atque spiritus turbatione. item bulimum magnitudine viarum stomachi atque ventris fieri sensit [sc. Asclepiades]; defectionem vero atque corporis fluxam et irregibilem laxitatem viarum inquit raritate fieri; item hydropismum perforatione carnis in parvam formulam viarum quae possit solita corporis nutrimenta inaquare.

And not all diseases are caused by an impaction of corpuscles, but certain ones, including *phrenitis, lethargia, pleuritis*, and violent fevers. *Solubiles* diseases are caused by a disturbance of liquids and *spiritus*. Asclepiades believes that bulimia comes about through the size of the pores in the gut and belly; fainting, bodily flux, and uncontrollable looseness come about through the openness of the pores. Again dropsy is caused by the boring of a small type of duct in the flesh, capable of turning the usual nutriment of the body into water.[62]

Solubilis is not attested as a technical medical term;[63] *turbatio* is an impossibly vague one, and could be applied to all sorts of things. But Caelius does use *solubilis* and *levis* side by side at *De morbis acutis* 1. 8 to describe a type of fever brought about by a generalized disturbance in the body rather than by a blockage of corpuscles in pores.[64] Drabkin in his note on this passage sees a reference simply to the relative gravity of different fevers, and he translates *solubilem atque levem* as 'less serious and more readily overcome'. In the light of general medical usage, this translation is probably quite justifiable.[65] But this would be to ignore altogether the role of the ἄναρμοι ὄγκοι or *soluta corpuscula*. If we consider the diseases which Caelius lists as attributable to 'disturbance', we find that some of them are hardly 'light' or 'easy to break up': like dropsy, *bulimia*, atony, and certain types of

---

[62] *De morbis acutis* 1. 107–8.

[63] It is rare in Latin before the 4th cent., and I do not wish to suggest that Caelius thought it diagnostic of Asclepiades' pathology. At *De morbis acutis* 2. 192 the word may be used in a quasi-technical sense, as far as Methodism is concerned: 'solubilis enim plurimus aer esse perspicitur' ('for a large amount of air is seen to have dilating properties'). There is no corresponding term for the *solubiles* in the related passage in Soranus, *Gynaecia* 3. 4, so my belief that it refers to a separate group of diseases is largely founded on the Caelius text.

[64] See above, n. 47.

[65] Note the use of *solvo* for a disease 'breaking up' at Caelius, *De morbis chronicis* 3. 100. Cf. also Celsus, *De medicina* 7. 3. 2 (*solutus morbus*); 7. 3. 4 (*solutis febribus*); 8. 4. 11 (*febricula soluta*); and the use at ps.-Galen, *Definitiones medicae* 19. 391 K (a Herophilean testimonium) of δύσλυτος of a chronic disease.

mental aberration. Caelius' extensive discussion of dropsy, in bk. 3 of the *De morbis chronicis*, gives some idea of how seriously that particular condition was regarded. The case is similar with *peripneumonia*; some 'followers of Asclepiades' described this condition as a 'parvi temporis solutio cum tumore atque febre' ('an acute *solutio* accompanied by inflammation and fever').[66] In all these cases, I would argue that the *solutio* is ultimately a *solutio* of the Asclepiadean corpuscles. Because of the role given to 'flux' in Methodist pathology, it is difficult to trace the Asclepiadean origins here; in fact it might well be impossible were it not for Caelius' constant hostility to the Asclepiadeans.

Asclepiadean corpuscular dissolution can be seen most clearly not in a disease aetiology, but in his account of digestion. For Asclepiades, digestion was a straightforward physical dissolution of food in the belly. His account was a controversial one; some witnesses claimed that Asclepiades denied any such thing as digestion—in the traditional sense of 'concoction'—but held that ingested food is broken down without undergoing any qualitative change. Caelius' account is the most detailed:

et neque ullam digestionem in nobis esse, sed solutionem ciborum in ventre fieri crudam et per singulas particulas corporis ire, ut per omnes tenuis vias penetrare videatur, quod appellavit λεπτομερές, sed nos intelligimus spiritum. et neque inquit ferventis qualitatis neque frigidae esse, nimiae suae tenuitatis causa, neque alium quemlibet sensum tactus habere, sed per vias receptaculorum nutrimenti nunc arteriam, nunc nervum vel venam vel carnem fieri.

Nor [says Asclepiades] does any 'digestion' take place in us; rather an undigested *solutio* of food appears in the belly and it passes through the various parts of the body, penetrating, so it seems, all the fine passages. He calls the product of this *solutio* λεπτομερές—we might call it *spiritus*. It has neither the quality of being hot nor that of being cold because of its extreme fineness, nor does it have any tangible quality. It travels through the parts which accept nutriment, becoming now artery, now nerve, vein, or flesh.[67]

---

[66] *De morbis acutis* 2. 142. (With the discussion of the *phthisica passio* in the same chapter.) It would be tempting to draw a distinction between a disease like this one, characterized solely, or mainly, by a *solutio*, and one caused by blockage but accompanied by a sympathetic dissolution.   [67] *De morbis acutis* 1. 113.

The λεπτομερές is identified here with the *solutio ciborum*; we may imagine that its origin lies in the characteristic weakness of the underlying corpuscles. It is without quality—that is to say without tangible quality—and this enables it to percolate through the body, reforming into different types of structure. Food in the belly remains 'raw' because it is not 'cooked' by the body.[68] There is a clear similarity between this type of explanation of digestion and the one offered by the renegade Erasistrateans in the *De naturalibus facultatibus*. According to the Anonymus Londinensis (25. 27–31), Erasistratus himself regarded the blood as nutriment. Asclepiadean blood was, of course, made up of corpuscles. (For the Erasistratean background, see bk. 2 of the *De naturalibus facultatibus* and the *Introductio sive medicus* 14. 697 K.) We might also note the omission of 'bone' from the Asclepiadean list of materials generated by the mobile nutriment in Caelius. Galen remarks that Erasistratus and his followers made the same omission. Incidentally, the Asclepiadean theory bears a passing resemblance to that of the Anonymus Londinensis himself, who at 25. 3 would have it that nutriment is taken up 'vapour-like' through the pores in the belly, and thence added to the body. It is on the question of the *faculty* involved that the Anonymus parts company with Asclepiades. All these theories may bear an ultimate relation to the Platonic account of ἀνάδοσις in the *Timaeus*, but this may tell us more about the general influence of the *Timaeus* in Hellenistic science and medicine than about any specific debts to Plato on the part of Asclepiades or Erasistratus. But to return to the Caelius text, how can we be so sure that the dissolution which takes place in the Asclepiadean belly is not simply one of food into its component corpuscles, rather than a *solutio* of the corpuscles themselves? Certainty is not possible, but given the likelihood that Asclepiades regarded solid matter as composed of large corpuscles, it is reasonable to imagine that some kind of

---

[68] This is the feature of the theory of digestion which attracts most attention. Vindicianus, *De semine* 8, p. 213 Wellmann (1901), notes 'vult enim [sc. Asclepiades] ex crudis fieri redhibitiones'; Anonymus Londinensis 25. 25 οὗτος (Ἀσκληπιάδης) γὰρ ἐξ ὠμῶν αὐτὸ μόνον λέγει γίνεσθαι τὴν ἀνάδοσιν. Cf. Celsus, *De medicina* 1, proem 20.

dissolution of the particles themselves will be required if the tenuous state of λεπτομέρεια is to be reached.

Let us consider drospy for a moment. As I noted earlier, Caelius preserves the following Asclepiadean aetiology: 'dropsy is caused by the boring of a small type of duct in the flesh, capable of turning the usual[69] nutriment of the body into water.'

It would be silly to speculate too far on exactly what lies behind this opaque sentence, but in the light of the Asclepiadean doctrine on digestion we could say that in a patient with dropsy the *solutio ciborum* is unable to reconstitute itself, but turns to liquid rather as urine is formed in the bladder. (In both cases, areas shot through with fine pores induce a condensation of vapours.)

There are a number of parallel situations in which liquid matter is generated in particular types of structure. Asclepiades seems to have explained the generation of bile in terms of the shape and position of the bile-ducts. This doctrine marks an important divergence from the humoral orthodoxy we find in Galen; in fact Galen is outraged by it.[70] Again, Asclepiades' explanation of how sputa collect in the lungs may be grounded upon the same idea.[71]

I can only guess at how this type of explanation might have been applied to the other diseases Caelius mentions at *De morbis acutis* 1. 107. But it does seem that these diseases were characterized by dissolution, by fragmentation of the corpuscles, or even by a change in the state of the pores themselves, rather than by impaction. The Methodists, as I shall argue in the final chapter, shifted their attention from the corpuscles to the pores.

All diseases, then, whether characterized by impaction or dissolution, could be explained ultimately in terms of the propensity of the ἄναρμοι ὄγκοι to shatter and then reconstitute

---

[69] I am tempted to read *soluta* for *solida* here. There is some textual confusion over the matter; see Bendz (1954), 137–9.

[70] *De naturalibus facultatibus* 2. 39–44 K. Asclepiades may owe something here to a doctor called Petron; see Anonymus Londinensis 20. 1–24.

[71] *De morbis acutis* 2. 98; if, indeed, this explanation should be attributed to Asclepiades. It is far from clear whom Caelius is talking about, and he ends up apparently endorsing the view himself.

themselves in different sizes. The dissolution is not one of compounds into corpuscles, but of corpuscles themselves into fragments of corpuscles. Void there may have been, but it is of no interest to our sources; it will come as no surprise to learn that Epicurus and Heraclides do not figure in the next chapter.

4

Before and After

In Chapter 2 I suggested that Asclepiades' theory of πρὸς τὸ λεπτομερὲς φορά (PTLP) should be seen as a reaction to the Erasistratean doctrine of πρὸς τὸ κενούμενον ἀκολουθία (PTKA). I noted a number of cases where Asclepiadean discussions of pathological and physiological phenomena could be connected with the corresponding Erasistratean accounts. The reaction, I argued there, was a complex one, and needs to be understood against the background of dissent amongst Erasistrateans who lived in the generations after their master. Now is the time to develop this a little further; the Asclepiadean corpuscular hypothesis can be related quite closely to certain aspects of Erasistratean physiology.

Asclepiades and Erasistratus are frequently mentioned together by our sources. In fact, Asclepiades is mentioned along with Erasistratus more often than he is with any other doctor, and certainly more often than with Heraclides or Epicurus. More telling still, perhaps, Asclepiades wrote several books attacking Erasistratus.[1] Connections like this on their own are hardly enough to suggest a solid link between them; Galen's assimilation of Asclepiades to Epicureanism should make us rather wary of taking evidence such as this at face value. In some

---

[1] Caelius, *De morbis acutis* 2. 173 'libris enim quos ad Erasistratum fecit et appellavit *Contradictorios*, dico, inquit [sc. Asclepiades] cardiacos non febricitare'. (Compare the Asclepiadean title *Ad Erasistratum* cited by Caelius at *De morbis chronicis* 5. 51.) Smith (1982) believes that the context here in Caelius suggests a work with therapeutic rather than theoretical concerns, but Caelius is never really interested in theory.

cases an implicit connection is made between the physical theories of the two individuals; more often, though, Galen mentions them together when he is attacking what he sees as their shared doctrine on the denial of innate faculties such as attraction, and of teleology. But the most notable case of connection comes at ps.-Galen, *Introductio sive medicus* 14. 699 K:

οἱ δὲ στερεὰ σώματα τὰ ἀρχικὰ καὶ στοιχειώδη ὑποθέμενοι, τά τε φύσει συνεστῶτα ἐκ τούτων καὶ τῶν νόσων τὰς αἰτίας ἐντεῦθεν λαμβάνουσιν, ὡς Ἐρασίστρατος καὶ Ἀσκληπιάδης.

Some people, like Erasistratus and Asclepiades, posit hard, primary, elemental bodies and understand the universe as made from them, as they do the causes of diseases.[2]

More significant still is a series of reports which suggest that Asclepiades and Erasistratus adopted a similar type of approach to pathological explanation. One of the fragments of Caelius Aurelianus' *Medicinales responsiones* opposes Erasistratus and Asclepiades in these terms:

uniformis est sanitas an varia cum extenditur et minuitur? secundum Asclepiadem et Erasistratum uniformis: nolunt enim summum sanitatis et quantitudinem esse. secundum Herophilum et Methodicos varia, cum extenditur et minuitur.

Is health one type of thing, or is it a variety of things, since it increases and diminishes? According to Asclepiades and Erasistratus health is one type of thing; for they do not accept that there is a peak and a measure of health. According to Herophilus and the Methodists, health is a variety of things, since it increases and diminishes.[3]

The idea of 'uniform' explanation is one which Galen's empiricist in the treatise *On Medical Experience* sees as particularly diagnostic of hardline dogmatism: 'the assertion of the dog-

---

[2] Compare Galen, *On Medical Experience* 24. 6, where the Asclepiadean and Erasistratean aetiologies of fever are mentioned together, even though Asclepiades and Erasistratus are not named, and 26. 7–8, where they are. At *De plenitudine* 7. 543 K Galen associates the Erasistratean doctrine on plethora with the Asclepiadean correlation between the amount of blood and the body's strength.

[3] Rose, *Anecdota* ii. 197. The text is uncertain. One MS reads *volunt* for *nolunt* and could conceivably be correct, depending on whether or not one takes 'cum extenditur et minuitur' as referring to 'uniformis est' or simply to 'varia'.

matists that by means of the *logos* they can bring into unity things which are utterly opposed to each other gives one cause for the greatest astonishment at the excellence of their intelligence.'[4] Galen makes the point again in connection with Asclepiades and Erasistratus:

(You [Erasistratus] declare that) the *logos* itself discovered (the case of) the phrenetic to be uniform. When he speaks thus, we answer, and say: what kind of *logos* is this? The *logos* of Erasistratos who operates with three kinds of vessels, or the *logos* of Asclepiades who operates with atoms and pores?[5]

The same point is made in a slightly different way by the author of the ps.-Galenic *Introductio sive medicus*:

According to Erasistratus and Asclepiades there is just one cause of all disease. For Erasistratus it is the transference (παρέμπτωσις) of blood into the arteries. For Asclepiades it is the impaction (ἔνστασις) of corpuscles in pores.[6]

The characterization of disease as uniform, reducible to one type of explanation, has a long history, going back at least as far as those sophistic Hippocratic treatises such as *On Breaths*.[7] As for Asclepiades, the implication is that he sought in some way to simplify the explanation of health and disease. (Several of the Asclepiadean *testimonia* cited above are doxographical, and therefore quite possibly simplified themselves, but the message seems clear.) But what about Erasistratus? How did his pathology and physiology operate? Unfortunately, no critical collection of the ancient Erasistratean *testimonia* yet exists,[8] and this account is a necessarily brief one.

---

[4] *On Medical Experience* 23. 2–3 (trans Walzer).

[5] Ibid. 27. 1–2 (trans. Walzer).

[6] 14. 728–9 K; see above, p. 93, for the text.

[7] The author of this work claims that: ἔστι δὲ μία πασέων νούσων καὶ ἰδέη καὶ αἰτίη (*On Breaths* 2).The influence of such unitary explanations in the Hippocratic corpus on later medical thought seems clear, and it is surely no coincidence that the Hippocrates of the Anonymus Londinensis papyrus is the Hippocrates of *On Breaths*, who explained disease in terms of the δύσροια πνεύματος (6. 14–18).

[8] The promised collection by von Staden is keenly awaited. For Erasistratus we are even more reliant on Galen than we are in the case of Asclepiades. While Galen basically found much to praise in Erasistratus' work (especially in anatomy), his attitude is often

The background is highly complex; the origins and development of a typical disease followed a pattern even more involved than the Asclepiadean. What follows is my reconstruction of the Erasistratean process leading to acute fever.

1. *The veins become congested with blood*

The first stage in an Erasistratean disease is generally termed 'plethora' ($\pi\lambda\eta\theta\acute{\omega}\rho\alpha$). LSJ interpret this as a 'repletion' of blood; more specifically, plethora for Erasistratus was a surfeit of blood in the veins. Galen offers a convenient characterization of the condition in the *De plenitudine*;[9] he outlines some of the antecedent circumstances which can increase the amount of blood in the veins,[10] but stresses most of all that *plethora* in the veins is simply the first stage in the process. Soranus confirms this: at *Gynaecia* 3. 4 he says that Erasistratus held plethora of blood to be an antecedent cause of all illness.

2. *Transference of blood from veins to arteries begins*

When more blood accumulates in the veins than they can naturally accommodate, and the 'filling up' ($\dot{\epsilon}\pi\acute{\iota}\sigma\alpha\xi\iota s$)[11] of the veins reaches a critical point, then a transference from veins to arteries, called $\pi\alpha\rho\acute{\epsilon}\mu\pi\tau\omega\sigma\iota s$, takes place. The unnatural presence of blood in the arteries marks the turning-point in the development of the morbid condition.[12] Blood in the arteries

hostile. The first two pages of Fuchs's 1892 Berlin dissertation on Erasistratus (1892*a*) contain an extraordinary (and highly entertaining) list of abusive epithets which Galen used against him.

[9] Galen, *De plenitudine* 7. 537–43 K; of $\pi\lambda\eta\theta\acute{\omega}\rho\alpha$, Galen says at 7. 537 K $o\ddot{\upsilon}\tau\omega s$ $\gamma\grave{\alpha}\rho$ $\alpha\dot{\upsilon}\tau\grave{o}s$ $\dot{o}\nu o\mu\acute{\alpha}\zeta\epsilon\iota$ [sc. Erasistratus] $\tau\grave{o}$ $\kappa\alpha\tau\grave{\alpha}$ $\tau\grave{\alpha}s$ $\phi\lambda\acute{\epsilon}\beta\alpha s$ $\pi\lambda\tilde{\eta}\theta os$.

[10] See also *De causis procatarcticis* 45. 18–23 Bardong, where over-eating is another factor mentioned as contributing to the $\pi\lambda\tilde{\eta}\theta os$ in the veins: 'oportet igitur ex repletione eorum que comeduntur et bibuntur in venis generari multitudinem, ab illa vero rursus paremptosim id est casum ad arterias fieri' (trans. Niccolo).

[11] *De plenitudine* 7. 541–2 K. I cannot decide if this was a technical Erasistratean term or not.

[12] *De plenitudine* 7. 541 K $\dot{\eta}$ $\gamma\grave{\alpha}\rho$ $\pi\lambda\eta\theta\acute{\omega}\rho\alpha$ $\pi\alpha\rho\epsilon\mu\pi\tau\acute{\omega}\sigma\epsilon\acute{\omega}s$ $\dot{\epsilon}\sigma\tau\iota\nu$ $\alpha\dot{\iota}\tau\acute{\iota}\alpha$ $\kappa\alpha\tau'$ $\alpha\dot{\upsilon}\tau\acute{o}\nu$, $\dot{\epsilon}\phi'$ $\tilde{\eta}$ $\phi\lambda\epsilon\gamma\mu o\nu\alpha\acute{\iota}$ $\tau\epsilon$ $\sigma\upsilon\nu\acute{\iota}\sigma\tau\alpha\nu\tau\alpha\iota$ $\kappa\alpha\grave{\iota}$ $\pi\upsilon\rho\epsilon\tau\grave{o}s$ $\dot{\epsilon}\pi\epsilon\tau\alpha\iota$. Cf. *De usu partium* 3. 493 K; *An in arteriis* 4. 713–14, 723–4 K; ps.-Galen, *Introductio sive medicus* 14. 692, 728 K; Anonymus Londinensis 26. 48 ff.

means that, under the terms of the Erasistratean theory, the natural passage of inspired pneuma is interrupted[13] and the resulting condition is called a σφήνωσις τῶν ἀρτηριῶν.[14] Erasistratus' pathology requires that liquids and pneuma in the healthy body be kept separate. The liquids belong in the veins, and pneuma in the arteries. Any disturbance of this balance, whether originating in the arteries or the veins, is dangerous.[15]

The transfusion of blood from vein to artery is made possible by the existence of tiny communicating pores in the walls of both arteries and veins. These pores are called ἀναστομώσεις or συναναστομώσεις.[16] Transfusion can be brought about by conditions other than plethora in the veins. In the classic Erasistratean illustration of why blood can sometimes be found in the arteries during dissection—a notorious problem for those like Erasistratus who believed it should not be there at all—it is clear that PTKA is invoked as the dynamic principle, sucking blood, so to speak, into the arteries through these interconnecting pores at the same time as the pneuma is evacuated from the arteries.[17]

3. *This transference gives rise to inflammation*

The interruption to the flow of pneuma through the heart is also associated with changes in the pulse, a rise in the pressure of

---

[13] See ch. 6 of Galen, *An in arteriis*, with Furley's revised arrangement in Furley and Wilkie (1984), esp. 4. 722 K.

[14] *De methodo medendi* 10. 101–2 K. Galen is not particularly interested in the details, but does suggest that an impaction (ἔμφραξις) in the veins is one factor which can lead to transference of blood into the arteries. Compare the [Asclepiadean] characterization of ἔνστασις at Cassius the Iatrosophist, *Problemata* 76: ἔνστασις δ᾽ ἐστιν ὄγκος ἐν λόγῳ θεωρητοῖς ἀραιώμασιν διὰ σφήνωσιν.

[15] For the idea that pneuma cannot safely coexist with liquids see Galen, *An in arteriis* 4. 722 K (in Furley and Wilkie's text): μὴ δυναμένου συνοικεῖν ἀμαχῶς πνεύματος ὑγρῷ.

[16] Various forms of the noun (and verb) are found; see Galen, *An in arteriis* 4. 709 K, *De usu partium* 3. 492 K, *De venae sectione adversus Erasistratum* 11. 154 K, and Anonymus Londinensis 26. 47 ff. The concept behind the term is discussed in detail by Galen in the *An in arteriis*; here, as usually, the word is associated with Erasistratus, although it appears in Celsus at *De medicina* 4. 11. 3, where it is attributed simply to *auctores medici*. At *De locis affectis* 8. 394 K Galen glosses the related term στόματα with πόροι.

[17] *De venae sectione adversus Erasistratum* 11. 154 K αὐτῆς δὲ τῆς παρεμπτώσεως αἰτίαν εἶναί φησιν [sc. Erasistratus] τὴν πρὸς τὸ κενούμενον ἀκολουθίαν. Cf. *An in arteriis* 4. 708–9 K; Anonymus Londinensis 27. 6 ff.

pneuma in the arteries, and a consequent forcing of blood to the sides of the arteries. This process ultimately brings about 'inflammation', variously called φλεγμονή, *tumor*, and *inflammatio* by our witnesses.[18] Erasistratean fever is a direct consequence of the heat of inflammation.[19]

The concept of blockage is important at several points in this Erasistratean process. Firstly, the inability of the veins to carry the amount of blood which they are called upon to carry causes the excess to ooze through the tiny pores in their walls into the arteries. The presence of blood in the arteries involves in itself a far more serious form of interruption to the natural flow of bodily fluids—in this case to the flow of vital pneuma. In a sense we could characterize disease for Erasistratus as a series of such impediments.

A number of parallels with the Asclepiadean account may be noted. The movement of vital fluids in Asclepiades' pathology is based on a concept not far removed from the one which informs the Erasistratean movement of pareptotic blood into the arteries. Second, if we include the tiny pores which allow communication between arteries and veins, along with the veins and arteries themselves, we might say that three types of vessels are implicated in the Erasistratean aetiology of disease. What we

---

[18] Celsus describes the process thus at *De medicina*, proem 15 'sanguis in eas venas quae spiritui accommodatae sunt transfunditur et inflammationem, quam Graeci φλεγμονή⟨ν⟩ nominant, excitat.' At *De usu partium* 3. 493 K Galen emphasizes that Erasistratean inflammation cannot occur unless blood is transfused into the arteries from the veins. This, he argues, is what makes παρέμπτωσις the corner-stone of Erasistratean pathology. See also *De methodo medendi* 10. 461 K.

[19] In addition to the Celsus passage mentioned in the previous note, see Galen, *De venae sectione adversus Erasistratum* 11. 235-6 K; ps.-Galen, *Introductio sive medicus* 14. 729 K; Apollophanes the Erasistratean *ap.* Caelius, *De morbis acutis* 2. 175-8. Ps.-Galen, *De historia philosophica* 131, illustrates the process with an analogy. The blood is like the sea. When no wind (i.e. pneuma) is blowing, it is quite, but when an unnatural wind blows up, it is disturbed. The blood then 'falls' into the arteries, and since it is hot, it heats the body. Hence fever for Erasistratus was a sign that this process is at work, rather than a disease in itself: Ἐρασίστρατος ὁρίζεται τὸν πυρετὸν κίνημα αἵματος παρεμπεπτωκότος εἰς τὰ τοῦ πνεύματος ἀγγεῖα ἀπροαιρέτως γινόμενον καθάπερ ἐπὶ τῆς θαλάττης, ὅταν μηδὲν αὐτὴν κινῇ πνεῦμα, ἠρεμεῖ. Although κίνημα (*motus* in Latin) is a common way of describing 'movement' into a disease, I cannot help comparing the prominence given to the *turbatio* of liquids and pneuma in Asclepiades' pathology.

find in Asclepiades under the general heading of 'pores' is the result of an amalgamation or reduction of these different types of Erasistratean vessels. Just as the veins and arteries are in a sense the centre of all affections for Erasistratus, so are the pores for Asclepiades. And so it continues. 'Impaction of corpuscles in pores' is, I would suggest, a generalization or simplification of the various types of morbid blockage and interruption in the Erasistratean account. Similarly, blood and pneuma, the mobile elements for Erasistratus, are very likely to be related to the ὑγρὰ καὶ πνεύματα of Asclepiades, especially in the light of those witnesses who associate both doctors with a rejection of one type of humoral pathology.[20] And Asclepiades, like Erasistratus, regarded plethora as an antecedent cause of disease.[21]

The Erasistratean element theory, involving the idea of a 'threefold web' of vein, artery, and nerve, can also be associated, if only indirectly, with the Asclepiadean theory of the ἄναρμοι ὄγκοι. The doxographers describe both theories together on several occasions, applying similar terminology to each.[22] I am not necessarily suggesting that Erasistratus held to a particulate theory of matter. But this parallel aligns Asclepiades with those later Erasistrateans who discussed the problem whether or not their master's elements were continuous, or themselves further divisible.[23] The theory of invisible emanations attributed to Erasistratus by the author of the Anonymus Londinensis papyrus may indicate the presence of a corpuscular theory of some kind; it certainly calls to mind the Asclepiadean doctrine

[20] This is a difficult area—I have already drawn attention to the variety in ancient humoral systems. Yet there do seem to be real parallels between the Erasistratean position as noted by ps.-Galen, *Introductio sive medicus* 14. 697 K (Erasistratus) παραλείπει τά τε ὑγρὰ καὶ τὰ πνεύματα and the Asclepiadean position as noted by Caelius Aurelianus at *De morbis acutis* 1. 112 'et [sc. Asclepiades dicit] non esse activas atque operantes causas aegritudinem in liquidis constitutas, quas synecticas vocant, sed esse antecedentes, quas Graeci procatarcticas apellant.'

[21] Something which even Galen approves at *Adversus Iulianum* 18A. 275-6 K, though this is probably because he wishes to get at the Methodists.

[22] Phrases such as κατὰ τὸ λόγῳ θεωρητόν, which are more often associated with Epicurean atomism. See e.g. Anonymus Londinensis 21. 23-9.

[23] See above, pp. 74-7.

on flux, though it would probably be rash to go further than that.[24]

All these observations can be summed up briefly in a table. In each case, I would suggest, Asclepiades can be seen to have 'simplified' a prior, Erasistratean account.

ERASISTRATUS	ASCLEPIADES
1. *Physiology*	
Body constituted out of three types of elemental vessel.	Body constituted out of one type of fragile corpuscle. Even the vessels made of them.
2. *Pathology*	
Rejection of qualitative humoral pathology.	Rejection of qualitative humoral pathology.
Veins/arteries the centre of affections.	Pores the centre of affections.
Movement of blood from veins to arteries results in blocking of arteries.	Movement of corpuscles to porous areas unable to accommodate them results in blockage.
Movement of blood to places where it should not be causes disease.	Movement of corpuscles to places where they should not be causes disease.
PTKA explains movement.	PTLP explains movement.
κίνημα a morbid condition.	*Turbatio* a morbid condition.

The terminology of the ἄναρμοι ὄγκοι may well come from Heraclides; the idea of a particulate theory, and especially the idea of void, may be related in some (distant) way to atomism. But Asclepiades' corpuscular hypothesis must be seen in its medical context if we are to make any sense of it. Even though Galen never states explicitly that Asclepiades' theory came from Erasistratus, there is good reason to suppose that it did, and that Galen knew as much.

---

[24] With the evidence for Asclepiades compare Anonymus Londinensis 22. 8–50, 30. 40 ff., and esp. 33. 44–5, where the author describes Erasistratus' famous proof that there are ἀποφοραὶ κατὰ τὸ λόγῳ θεωρητόν from a bird.

ASCLEPIADES AND THE METHODISTS

There has never been much doubt that Asclepiades is to be associated in some way with the development of Methodist medicine.[25] There were a few, somewhat neurotic, figures (like Cassius the Iatrosophist) who more or less closely followed Asclepiades' teaching well into the first century AD, but they are not particularly important. Of course, we have seen how much the Methodists hated Asclepiades—and Soranus himself barely mentions him. But close relationships often breed this sort of thing. To give the Methodists their due, if we set the corpuscular hypothesis against the generalized antagonism towards speculation and theory which characterizes the thinking of Soranus, Caelius Aurelianus, and the other Methodists, it is not immediately easy to see exactly where Asclepiades fits in. And if Galen presents Methodism as Asclepiadeanism in all but name, we have learnt to be careful about what he says. When we come to Methodism, we come to the immediate, contemporary focus of Galen's polemic. In his propaideutic work *De sectis ad introducendos* he manipulates his account of the medical sects so as to give the impression that Methodism alone is quite absurd and irrational, even when set against the theory of Asclepiades.[26] So what happened to Asclepiades' theory? Did it simply vanish without trace, or did the Methodists take it on, and then cover their tracks? In this section I shall again concentrate on Caelius

---

[25] Different aspects of the links between Asclepiades and the Methodists have been dealt with by many scholars; notable treatments include those by Raynaud (1862), 6; Wellmann (1913), 64 ff.; Meyer-Steineg (1916); Edelstein, PW suppl. VI s.v. 'Methodiker'=(1967), 173-91. Questions relating to Methodist epistemology and its philosophical affiliations have been tackled in recent years by Viano (1981), 650; Marelli (1981), 663; Frede (1982); Lloyd (1983), 182-200, (1987), 158-71. A surprising amount of modern work on the subject was anticipated in the voluminous and undervalued *De medicina methodica* of Prosper Alpinus, which appeared in Padua in 1611.

[26] He does this by cleverly assimilating Dogmatists and Empiricists (and even Asclepiades), suggesting that they all believe in much the same god, in spite of their doctrinal differences, and then setting them all against the Methodists, who are the only real medical heathens. For still more extreme cases, see *De methodo medendi* 10. 124, 841-3 K; *De simplicium medicamentorum temperamentis ac facultatibus* 11. 783 K.

Aurelianus as the typical Methodist, because he is the most important extant Methodist source for Asclepiades.

A few initial caveats may be in order. Methodism was not a homogeneous system, and our first-hand knowledge of it does not extend very far beyond what we can see in Caelius and Soranus.[27] The method of the Methodists was essentially a method of treatment, and while all Methodists seem to have shunned theoretical speculation, they did so to varying degrees. Thessalus, for instance, is frequently attacked by Caelius for being too theoretical. One basic doctrine does, however, seem to have been shared by them all: physical affections can be divided into three general types, or 'communities'. Put very briefly, one type involved a state of 'stricture', another 'flux', and another a mixture of the two.[28] There are no extant Methodist sources which discuss the establishment of these communities, but it is safe to say that they were general conditions which prevailed throughout the diseased body. Their importance seems not to have lain in their power to explain what was going on, but in their indicative ('endeictic') power to suggest the correct course of treatment. A patient afflicted by a state of stricture, for example, might require a treatment which loosened the stricture, and so on. Most important, the Methodist doctor did not really *need* to know anything about the theoretical background to these communities—quite simply, they were phenomenal, and there for anyone to see. The doctor's first task was to recognize the prevailing state; he could theorize about it if he really wanted to, but only in his own time.

One of the first post-classical scholars to come close to what I believe is the correct understanding of the link between Ascle-

---

[27] There are many other texts generally held to be Methodist—notably those printed by Fuchs (1894*a*; 1903)—but more work needs to be done on their true affiliations. These particular texts share the basic arrangement found in Caelius' work, but the doctrines discussed in them with apparent approval are not always what we might think to be Methodist. Rubinstein (1985) argues that there were as many types of Methodism as there were Methodists; certainly the distinction between 'old' and 'new' Methodism first advanced by Wellmann (1895; 1922) needs to be reconsidered.

[28] See e.g. Celsus, *De medicina* 1, proem 54-5: the Methodists posited three *genera* of diseases, *unum adstrictum, alterum fluens, tertium mixtum*. And, in Greek, Dionysius the Methodist *ap.* Soranus, *Gynaecia* 1. 29. 3.

piades and the Methodists was Prosper Alpinus. He argued[29] that the Methodist states could be derived from the ideas which lay behind the *meatus vel clausus vel fluxus*, the 'closed or open flow', of the Asclepiadean corpuscles. Yet because Alpinus believed that Asclepiades was an Epicurean atomist, he did not suggest that the *meatus fluxus* could perhaps be the result of some change in the physical structure of the particles themselves. Wellmann revived the idea at the beginning of this century. In a footnote to an article which sought mainly to argue for Methodist influence in the second half of the Anonymus Londinensis papyrus, he drew several parallels between, on the one hand, Asclepiadean impaction and the Methodist state of stricture, and, on the other, between the Asclepiadean *concursus corpusculorum* and Methodist flux.[30]

There is one crucial problem with Wellmann's parallel. *Concursus corpusculorum* is Caelius' synonym for *statio* or *obtrusio*; it bears no obvious relation to any concept of 'flow'. The movement of the Asclepiadean corpuscles, implicit in the idea of a *concursus*, is simply a movement which leads directly to blockage;[31] it would be more plausible to suggest that the Methodist equivalent to this would again be the state of stricture. How should this problem be resolved?

Connections between the two main Asclepiadean aetiologies and Methodist communities are evident on two levels—that of terminology, and, in my view, that of actual doctrine. I am not suggesting that the Methodists were closet Asclepiadeans, as

---

[29] pp. 8–9.

[30] Wellmann (1922), 398 n. 3, presented the parallel in these terms: 'die Methodiker an die Stelle des asklepiadeischen Terminus "statio corpusculorum" (ἔνστασις τῶν ὄγκων) πύκνωσις setzten und für "concursus corpusculorum" ῥύσις.' Wellmann had earlier gone further still (1905), claiming (p. 600) that the ἔνστασις, occurring in the anonymus Paris MS of the work he identified as Herodotus' treatise on acute diseases, was actually a Methodist term. In all this he was broadly followed by Edelstein (1967), 187: 'Themison modified the Asclepiadean system by turning the atoms and pores into common conditions.'

[31] Caelius uses *concursus corpusculorum* as equivalent to *obtrusio* in the very passage with which Wellmann supported his argument: *De morbis acutis* 3. 188 'cholera, inquit [sc. Asclepiades], est humoris fluor, celer ac parvi temporis, ventris atque intestinorum ex concursu sive obtrusione corpusculorum'; cf. ibid. 220 'ex corpusculorum concursu sive conventu, quem *enstasim* appellavit [sc. Asclepiades]'.

Galen would have us believe, nor that they took over the corpuscular hypothesis in any consciously systematic way. Consider
the following table:

ASCLEPIADES	METHODISTS
ἔνστασις	στέγνωσις[32]
statio[33]	
obtrusio	strictura[34]
coacervatio	adstrictus[35]

The case of στέγνωσις offers a convenient illustration of the
comparison. στεγνόω can carry a sense of pathological blockage
or contraction in medical contexts.[36] Similarly, *strictus* (from
stringo) can have the sense 'closely packed', 'dense'.[37] For the
Methodists, such terms relate not to the nature of the 'close-
packing', but to the state which results. It is inevitable, though,
that the use of such terms will presuppose certain assumptions
about the origins and nature of the state. These assumptions are
fairly evidently Asclepiadean. Just as I have noted some signs of a
generalization of Erasistratean pathology in Asclepiades' system,
here we have a case where a specific malfunction in the body,
located in a specific place, is generalized to the point where it
becomes an overall state. Each Asclepiadean term relates to the
actions of the corpuscles. The corresponding Methodist general-
izations suggest the resultant situation without referring to its
aetiology at all.

---

[32] For the Methodist term see e.g. Soranus, *Gynaecia* 1. 28. 2–3; Galen, *De sectis* 1.
79–80 K.

[33] e.g. Soranus, *Gynaecia* 3. 28. 1, 49. 3; Galen, *De methodo medendi* 10. 20 K (of
Thessalus).

[34] e.g. Caelius, *De morbis acutis* 1. 52.

[35] e.g. Celsus, *De medicina* 1, proem 55.

[36] Hence Oribasius' rather general characterization of the concept at *Synopsis* 5. 16. 1,
which I mentioned in the last chapter: ἡ δὲ στέγνωσις γίνεται μὲν δι' ἔμφραξιν ἢ
πύκνωσιν τῶν πόρων, 'stegnosis comes about through blockage or thickening of the
pores'. Oribasius follows this with a list of the signs by which it can be recognized. Cf.
Galen, *De sanitate tuenda* 6. 218–19 K. The state will be one of tension: the idea is used for
constipation at Alexander, *Problemata* 1.

[37] See Celsus, *De medicina* 4. 1. 10 (of the *omentum*); Scribonius Largus, *Composi-
tiones* 45.

What of the Methodist flux? This is a more difficult case, largely because our evidence for the Asclepiadean diseases not caused by blockage is so slight. To recall the conclusion reached in the first section of this chapter, it is the 'weakness' of the Asclepiadean corpuscles which underpins the whole of the corpuscular hypothesis. The corpuscles may congregate in certain parts of the body, causing impaction, or may break up and rush away, giving rise to various types of unnatural flow. In the case of the latter group of affections, we have seen how the *solubulis* nature of the particles explains their behaviour. There is a strong case, I believe, for supposing that the fragility of the corpuscles ultimately lies behind the Methodist community of flux. Once again there are interesting terminological parallels:

ASCLEPIADES	METHODISTS
Solutio of ἄναρμοι ὄγκοι	*solutio*[38]
or *corpuscula soluta*[39]	ῥοῶδες, ῥύσις[40]
	fluens[41]

One of the many mysteries surrounding the Asclepiadean theory was its disappearance from later medicine. It is in this aspect of Methodist doctrine that we could well see both what happened to the ἄναρμοι ὄγκοι and to the ideas which they represented. The state which the Methodists call 'flux' has its origins in a *solutio* of the body; for them the whole body, not the corpuscles, is ἄναρμος. The Methodists covered their tracks with great efficiency. The 'mixed' disease state is not explicitly discussed in what remains of Soranus, but it figures prominently

---

[38] e.g. Caelius, *De morbis acutis* 1. 60. Caelius equates *solutio* with *raritas viarum* at *De morbis acutis* 2. 177. This is because of a general shift of attention on the part of the Methodists from the Asclepiadean corpuscles to the pores.

[39] e.g. Soranus, *Gynaecia* 1. 29. 3.

[40] In his discussion of Methodism at *Pyrrhoniae hypotyposes* 1. 238, Sextus opposes στέγνωσις to χαύνωσις. χαυνόω shares a number of senses with *solvo* in Latin and with ἄναρμος in Greek. Similarly, Plutarch, *Sertorius* 771 B remarks that the soil around the Tagonius is ὑπὸ χαυνότητος εὔθρυπτος. The author of the *Geoponica* contrasts ἕνωσις and χαύνωσις at 10. 75. 17 (where he is using the terms for strong and weak grafts on to plants).

[41] e.g. Celsus, *De medicina* 1, proem 55.

in Caelius Aurelianus. It corresponds most naturally to the Asclepiadean cases where there is one immediate cause, and a supervening, sympathetic affection. Diarrhoea is perhaps a good example.

These parallels between Asclepiadean theory and Methodist non-theory can also be seen in similar characterizations of disease. Consider the following table:

ASCLEPIADES	METHODISTS
1. *Phrenitis*	
'Statio corpusculorum in cerebri membranis frequenter sine consensu.'	A disease either of stricture or of a mixture of stricture and *solutio*.[42]
2. *Pleuritis*	
Fluor, accompanied by *inflammatio*.	A disease either of stricture or of a mixture of stricture and *solutio*.[43]
3. *Pneumonia*	
Solutio with *inflammatio*	A disease either of stricture or of a mixture of stricture and *solutio*.[44]
4. *Cardiaca passio*	
Tumor in the heart, brought about by a piling up of corpuscles.	A disease of *solutio*.[45]
5. *Synanche*	
Humoris fluor	A disease of stricture.[46]
6. *Tetanus*	
Extentio corporum . . .	A disease of stricture.[47]

This table at first sight seems unhelpful. Four of the six cases do not seem to show parallels at all. But the parallels I would

---

[42] Asclepiades: Caelius, *De morbis acutis* 1. 6; Methodists: ibid. 1. 52, 142.
[43] Asclepiades: ibid. 2. 89; Methodists: ibid. 2. 90.
[44] Asclepiades: ibid. 2. 142; Methodists: ibid.
[45] Asclepiades: ibid. 2. 163, 174; Methodists: ibid. 2. 163.
[46] Asclepiades: ibid. 3. 5; Methodists: ibid. 3. 5, 10.
[47] Asclepiades: ibid. 3. 63; Methodists: ibid. 2. 64–5.

note are not on the level of specific doctrine, rather on the level of general approach. In fact, the links between Asclepiades and the Methodists here are quite close. First *phrenitis*: by positing two types of the disease, one involving stricture alone and the other a mixed state of stricture and flux, Caelius and Soranus seem to be looking back to the Asclepiadean concept of this disease as one which may or may not involve a secondary affection. The secondary affection, according to Galen, was the 'flux' in the belly; it becomes the *solutio*. At *De morbis acutis* 1. 14 Caelius notes: 'neque nunc obstrusio causa est phreniticae passionis, sicut in aliis ostendimus, neque causam debuit pro effectu accipere, neque in membranis consequentem obtrusionem fieri, sed in sensibus, hoc est in sensualibus viis' ('we have shown elsewhere that obstruction is not the cause of the phrenitic disease; nor should we take the cause for the effect. Nor indeed does this obstruction come about in the (cerebral) membranes, but in the "senses", that is to say the sensory passages'). It is not clear from this whether he means categorically that blockage is not the cause of the disease. The second part of the sentence suggests that he is not ruling out the Asclepiadean aetiology.

When Caelius first sets out his understanding of phrenitis, he mentions only the visible signs of the disease, but at *De morbis acutis* 1. 52 makes it clear that he is ultimately not all that far away from Asclepiades' own position:

nos vero aliam dicimus esse ex strictura, aliam ex complexione stricturae atque solutionis. est enim verum ita discernere, ut non accidentium diversitas passionis differentias ostendat, sed generalis quaedam ac necessaria designatio, quae fiet ut supra diximus ex principalibus passionibus, unde etiam curationum ratio sumatur.

In fact, we maintain that one type (of phrenitis) is characterized by stricture and another by a mixture of stricture and *solutio*. For it is correct to distinguish them in this way so that a general and necessary indication points to differences between diseases, rather than the variety of symptoms. The general and necessary indication is derived, as I have said, from the basic affections [i.e. the communities], and from this comes our reasoning about cures.

What of *pleuritis?* This case can be explained by Caelius' note at *De morbis acutis* 2. 89:

item Asclepiades libro Diffinitionum pleuritim dicit esse humoris fluorem temporis parvi atque celerem interiorum lateris partium cum febre atque tumore, sed erat melius tumorem dicere atque humoris fluorem: etenim quoties ex complexa passione conficitur, tumorem perspicimus superare.

Again, Asclepiades says in his work on Definitions that 'pleuritis' is an acute flow of liquid over a short period of time in the internal lateral parts [i.e. of the chest], accompanied by fever and swelling. But it would have been better to call it a swelling *and* a flow of liquid. For in cases where [pleuritis] is made up out of a mixed disease, we see that the swelling is more significant.

Caelius' objection to the Asclepiadean characterization of the disease does not take him outside the theoretical framework of the corpuscular hypothesis. Rather, he simply takes exception to the emphasis of the Asclepiadean account and suggests that the swelling in *pleuritis* is somehow the more significant sign of the underlying condition. *Tumor* for Asclepiades is the result of an impaction,[48] so we should expect the Methodist to translate this to a state of stricture. Caelius does not deny that the *fluor* takes place; in fact he admits as much by conceding that the disease can sometimes represent a 'mixed' community.

The case is much the same with *pneumonia*. Asclepiades claims that this affection is an acute *solutio* in the lungs, accompanied by fever and swelling. Caelius, who claims explicitly to be quoting Soranus at this point, replies that 'we should not call this disease a flow (*rheumatismum*) or a *solutio* accompanied by *tumor*, but perhaps a *tumor* accompanied by a flow. Lesser details should give way to what is more important'.[49]

---

[48] This can be inferred from the characterization of the *cardiaca passio* at *De morbis acutis* 2. 163 'item Asclepiadis sectatores aiunt tumorem secundum cor corpusculorum coacervatione sive obtrusione effectum'. Cf. ibid. 2. 228, where Caelius prescribes *laxativa cataplasmata* as the appropriate treatment for *tumor*. The expression *solutis inflammationibus* (Celsus, *De medicina* 5. 26. 20 F) may also reflect the polarity between the two states, even though doctors can talk of diseases 'breaking up' without necessarily pointing to any particular theory.

[49] *De morbis acutis* 2. 142 'huic Soranus occurrens diffinitioni ait: "cum semper strictura

Hence the Methodist characterization appears at first to be the exact opposite of the Asclepiadean, although it operates in a similar way.

Likewise *synanche*. The *cardiaca passio* is slightly different. Caelius, again explicitly quoting Soranus at *De morbis acutis* 2. 163, denies that there is any sign of inflammation in the affection:

Soranus vero, cuius haec sunt, tumoris inquit signum nullum subesse quod in cardiacis videatur. item cor pati non valde plurimis probabile videtur. sed ait cardiacam esse passionem solutionem celerem atque acutam qua disici corpora per omnes viarum particulas apprehendit.

But Soranus, whose doctrine this is, says that there is no underlying sign of *tumor* to be seen in cases of the *cardiaca passio*. Furthermore, he says, it is not at all widely held that the heart is the affected part. Rather, he claims, the *cardiaca passio* is a swift and acute *solutio*. By this he understands that bodies [*sic*] are scattered throughout all the individual pores.[50]

His reasons are complex, and they are set out at 2. 173; Soranus and Caelius insist that the fevers, which Asclepiades claims may sometimes accompany the disease, can equally arise from a state of *solutio* as from one of stricture.[51] Here again we find these two Methodists discussing a problem in Asclepiadean intellectual categories; their disagreement with Asclepiades is beside the point, since it is couched in Asclepiadean terms. In fact the whole of Soranus' categorization of disease can only really be understood against the background of Asclepiades' theory.

There is one particularly striking difference between the Asclepiadean and the Methodist lists of diseases. In every Asclepiadean case, a specific *locus* for the affection is provided. This

in his obtineatur, non oportuisse passionem rhematismum vel solutionem dici cum tumore, sed forte tumorem cum rheumatismo: etenim parva maioribus postponuntur".'

[50] Drabkin translates *corpora* as 'bodily substance', and he may be right. I could not resist at least the suggestion that the word should be translated strictly. *Disici* coud also mean 'broken up', in which case we would have a nice example of Asclepiadean corpuscles in Soranus.

[51] *De morbis acutis* 2. 176-7.

locus affectus gives the doctor important information about the right therapy to follow. Caelius actively avoids this. To take the example of phrenitis: an affection characterized for Asclepiades by a blockage centred in the cerebral membranes becomes for Caelius a state of stricture which pervades the *whole body*. The *solutio* which may accompany Asclepiadean phrenitis is located in the stomach; the Methodist would make no such attempt at localization. If there is a flux in phrenitis, it is a general state.[52]

The pathology of the Methodist communities, then, can be seen as a generalized reduction of the Asclepiadean system. If my conclusions about Erasistratus and Asclepiades are correct, then Asclepiades himself was in all probability generalizing some of the principles of Erasistratean pathology, reducing the τριπλοκία to one kind of particle. Abandoning the idea of a *locus affectus*, explicitly at least, was a key feature of the Methodist development of the corpuscular hypothesis. This raises one serious question: why did the Methodists, with their disdain for speculation, ground their pathology on such a speculative theory? To be sure, no Methodist admits as much, but the case seems a strong one.

There are few clues to the origins of Methodism beyond what we have here with Asclepiades. But, as I have suggested, there does seem to have been a shift of attention to the pores—and pores, after all, are visible structures in the body, even to the naked eye. This may well have paved the way for the suggestion that something which takes place in a part of the body inevitably affects the whole because of the vast network of interconnecting pores. Caelius says that some Methodists corrected the Ascle-

---

[52] See *De morbis acutis* 1. 58–65; cf. 2. 183–5, dealing with the *cardiaca passio*, which Erasistratus and Asclepiades locate in the heart. The Methodists reply: 'nos vero cum Sorani iudicio totum videmus corpus in solutionem laxari, totum necessario pati accipimus. et neque valde nobis de praepatienti loco certandum est, ne in occulta quaestione versemur.' In spite of this, Caelius/Soranus frequently do specify a *locus affectus*: note e.g. *De morbis acutis* 2. 89 and the characterization of 'pleurisy': 'est igitur secundum Soranum pleuritis dolor vehemens interiorum lateris partium cum febribus acutis et tussicula qua variae qualitatis liquor excluditur.' Note also Caelius' attempts to distinguish 'spasm' and 'tetanus' at *De morbis acutis* 3. 61; between the *cardiaca passio* and certain diseases of the oesophagus at ibid. 2. 184. This raises important questions of conflict between Methodist theory and practice which have not yet been sufficiently addressed.

piadean definition of *cholera* by substituting *raritas viarum* for *concursus corpusculorum* at *De morbis acutis* 3. 189.[53] It was this passage which led Wellmann to believe that these were terms of roughly equivalent sense. This is not so; Caelius uses *raritas* of an increase in the size of the pores (cf. *De morbis acutis* 2. 177). He is unhappy here simply because the question involves doctors in unnecessary aetiology. The implication is that a Methodist is free to believe what he wants in private, but should keep it to himself.[54]

Neither Caelius nor Soranus directly relates a belief in the existence of invisible pores to the establishment of their disease communities, but some kind of relationship is certainly there.[55] Galen is keen to notice it wherever he can. He is clearly being hostile when he characterizes the Methodist disease states as $τὸ$ $μὲν\ εὐρύτης\ τις\ πόρων,\ τὸ\ δὲ\ στέγνωσις,$[56] and elsewhere he tries to remind them of their theoretical, Asclepiadean inheritance.[57] It would be foolish to write off Galen's testimony about the Methodists simply because it overestimates their acceptance of theory. Themison in particular is accused by Caelius and Soranus themselves of accepting too much to the Asclepiadean theory *and* being prepared to admit as much. Themison, says Caelius several times, was almost an Asclepiadean.[58] But I would

---

[53] 'item aliqui nostrorum tradiderunt eandem diffinitionem, solum concursum corpusculorum detrahentes atque viarum raritatem adicientes. nos autem superfluum fuisse causas passionis dicere iudicamus, cum sit necessarium id quod ex causis conficitur edocere.'

[54] And so Galen snipes at the Methodist physician Julian for 'secretly indulging in speculation' at *Adversus Iulianum* 17A. 255-7 K.

[55] On Soranus' implicit acceptance of invisible pores see *Gynaecia* 1. 35. 4-6 and Lloyd (1983), 192-3. [56] Galen, *De morborum differentiis* 6. 842 K.

[57] These cases usually reflect Galen's concern about the threat posed by Methodism in his own day. The prominent Methodist Julian was still alive, it would seem (*De methodo medendi* 10. 53 K). Apart from the extended discussion in the *De morborum differentiis* 6. 842 ff. K, see also *De causis morborum* 7. 1-2 K, and 32-3 K, where the Methodists are effectively assimilated to Asclepiadeanism. Also *De plenitudine* 7. 514-15 K; *De simplicium medicamentorum temperamentis ac facultatibus* 11. 783 K. When Galen speaks (at 11. 423 K) of 'leaders of the sects' who say that air is made up of fine corpuscles, he means Methodists, not strict followers of Asclepiades. Similarly, at *De naturalibus facultatibus* 2. 80 K he says $λῆρος\ οὖν\ μακρὸς\ ἅπασα\ πόρων\ ὑπόθεσις\ εἰς\ φυσικὴν\ ἐνέργειαν$, attacking Erasistrateans, Asclepiadeans, and Methodists all at once.

[58] For Caelius' criticisms of his Methodist forebears (Themison in this case) see *De morbis acutis* 1. 155, 2. 232 ff., 3. 29, *De morbis chronicis* 1. 48, 50, etc. Note, however, that

suggest that the concepts of ποροποιΐα and μετασύγκρισις (which embody the Methodist idea that changing the state of the pores is the first stage of a successful treatment) are clearly grounded in what Asclepiades had to say about the conditions which give rise to blockage.[59]

There is little doubt that Galen exaggerates the closeness of Asclepiades and the Methodists. This, and the fact that the Methodists say very little about the connection themselves, can make Asclepiades seem more 'Methodist' than I believe he really was. There is a considerable difference between characterizing health as a 'balance of the elements', as Asclepiades almost certainly did, and as a 'balance of the pores'. Frede[60] does not, in my view, distinguish sufficiently between Asclepiadean and Methodist pathology when he summarizes the Asclepiadean position in the following terms: Asclepiades 'explained many illnesses as owing to the constriction of . . . invisible pores, some owing to an excessive flow through them'. It is the role of the corpuscles, and not the pores, which seems to have been underlined by Asclepiades.

But why did the Methodists adopt this theory at all? They criticized Asclepiades for falling into the trap of speculating about the nature of man. Similar charges could be levelled at them. Their practice of naming affections, for instance, seems very odd, for doctors whose pathology explicitly rejected the need for disease entities. If we accept Caelius' explanation for his

Caelius may have regarded Themison as a late convert to Methodism: 'haec nunc Themison phreniticis curandis ordinavit, sed post ex methodica secta multa bona contulit medicinae' (*De morbis acutis* 1. 165). For Themison's theory and its supposed similarity to Asclepiades', see *De morbis acutis* 2. 52.

[59] These concepts seem to have figured most strongly in Thessalus' pathology—see Galen, *De methodo medendi* 10. 267–8 K, *De simplicium medicamentorum temperamentis ac facultatibus* 11. 782 K. The term ποροποιΐα does not figure in Soranus' *Gynaecia*. For Soranus and Caelius, μετασύγκρισις is used purely in the context of therapeutics, where the concept informs the correct selection of drugs to alleviate a given disease state: see *Gynaecia* 3. 16. 3–4, Caelius, *De morbis chronicis* 4. 16–18, *Gynaecia* 3. 15. 1–4, Caelius, *De morbis chronicis* 1. 24, *Gynaecia* 4. 39. 3–4, Caelius, *De morbis acutis* 2. 222–3.

[60] Frede (1982), 16. Smith (1979), 223, does much the same, saying that Asclepiades 'considered that too much constriction causes inflammation . . . while too much looseness allows loss of vital material'. There is no ancient evidence supporting this. Frede (1985) is much closer to my own position.

use of names for diseases—a simple matter of convenience—we must ask what price he had to pay for this convenience in terms of the integrity of Methodism. Yet this may well be to miss the main point about this sect: it sought to shut the door on theory altogether by basing itself on a system so speculative and so generally based that it left practitioners to get on with the business of treating their patients. The motivations of Asclepiades may not have been wholly dissimilar.

5
Asclepiades?

Now the Asclepiadean corpuscular hypothesis can be seen for what it was.

What it was not was Epicurean. To the extent that it stands in a tradition of 'monistic' theories, the theory might broadly be termed 'atomistic', but this would be to misrepresent it seriously. Galen is the only ancient witness to call Asclepiades an Epicurean, but whatever Galen says, the admission of breakable particles marks a crucial and inescapable departure from the tradition of Epicurean atomism. Galen's refutations of Asclepiades (and Epicurus for that matter) owe more to the arguments in the Hippocratic treatise *On the Nature of Man* than to a close acquaintance with the details of Hellenistic physical theories. So what about void? It does seem likely that Asclepiades envisaged some kind of role for Abderite-type void in his physics, even if it figures in no surviving Asclepiadean accounts of natural phenomena. My own suspicion is that Galen felt that this was enough: the theories of both Epicurus and Asclepiades committed them both to a heretical stance on the operation of teleology in nature. Atomism, after all, was the prominent anti-teleological system in antiquity, and much of Galen's assimilation of Epicurus and Asclepiades takes place in his two great hymns to teleology, the *De naturalibus facultatibus* and the *De usu partium*. But most important of all, Galen tends to mix in his Asclepiadean pot all the theories he dislikes most, and before pouring the brew over the Methodists. It is rhetoric, rather than philosophical history, that Galen offers us.

What the theory probably was not was Heraclidean. I have

said next to nothing about Heraclides of Pontus. This will seem surprising to those scholars who have assumed that Asclepiades is a transparent source for the Heraclidean theory. We know nothing at all about Heraclides' ἄναρμοι ὄγκοι. Heraclides may well have had some kind of corpuscular theory—after all, Plato had hinted at such a theory in the *Timaeus*, and Galen's commentary on that work makes it plain that later members of the Academy were interested in developing the physical doctrines of Plato. But Asclepiades must be understood in his own medical tradition, and close intellectual links with Erasistratus look far more likely than anything else. Perhaps Heraclides could be included in the equation, by way of Erasistratus. Similarly, there may well be connections with Stoicism waiting to be examined.

The corpuscular hypothesis of Asclepiades marks a notable attempt at paring down the multiplicity of explanations of physiological and pathological phenomena. Not only does this type of activity lie in a well-attested Hippocratic tradition, but in one whose importance in the development of Hellenistic medicine seems often to be overlooked. The sophistic show-pieces like *On Breaths* did have their own intellectual progeny. Our picture of Hippocratism is so coloured by Galen that it is easy to forget the importance of this tradition. In Asclepiades' corpuscular hypothesis, we have one stage in a reductionist process which may have begun with Erasistratus. What Asclepiades adopted was a theory which posited a uniform type of explanation for *all* affections in the body. A corpuscular theory, relying on the aleatoric movement and combination of particles, was an ideal way of defusing the problem of reconciling support for a hypothesis with the difficulties raised by the failure of hypotheses in prognosis. The Asclepiadean hypothesis could explain its own failure; to judge by the development of Methodism, it was ultimately an anti-theoretical theory.

The way is now open for a reassessment of Asclepiades' role in the development of Hellenistic medical epistemology. Such an inquiry will need to underline the importance of studying medical philosophers in their own intellectual surroundings. A closer study of the interactions between Stoics and Epicureans is also needed. It has recently been suggested that Hero of Alexandria shows signs of Stoic influence. Could it be that Asclepiades was a continuum theorist *manqué*? The story is not over yet.

Bibliography

1. Ancient Texts

These are bibliographical details of some of the less well-known texts I have cited, together with the editions of the major witnesses used in this study.

AETIUS DOXOGRAPHUS, *Placita philosophorum*, in *Dox.* 273–44.

—— *Aetius Arabus: Die Vorsokratiker in arabischer Überlieferung*, ed. H. Daiber (Wiesbaden, 1980).

AGNELLUS OF RAVENNA, *Commentarium in De sectis Galeni*, in *Lectures on Galen's* de Sectis (Latin Text and Translation by Seminar Classics 609, State University of New York at Buffalo, Department of Classics; *Arethusa* Monograph 8; New York, 1981).

PS.-ALEXANDER, *Problemata*, in J. L. Ideler, *Physici et medici Graeci minores*, i (Berlin, 1841; repr. Amsterdam, 1963).

ANONYMUS LONDINENSIS, *Iatrica*, ed. H. Diels, *Anonymi Londinensis ex Aristotelis iatricis Menoniis et aliis medicis eclogae* (Supplementum Aristotelicum, 3/1; Berlin, 1893).

—— *The Medical Writings of Anonymus Londinensis*, ed. W. H. S. Jones (Cambridge, 1947).

CAELIUS AURELIANUS, *On Acute Diseases and on Chronic Diseases*, ed. and trans. I. E. Drabkin (Chicago, 1950).

—— *Tardarum passionum libri V*, ed. J. Sichart (Basle, 1529). [S]

—— *Liber celerum vel acutarum passionum, qua licuit diligentia recognitus, atque nunc primum in lucem aeditus*, ed. J. W. von Andernach (Paris, 1533). [G]

—— *Caelii Aureliani de acutis morbis libri III; de diuturnis libri V ad fidem exemplaris manuscripti castigati* . . ., ed. G. Roville (Lyons, 1567). [R]

—— *De morbis acutis et chronicis libri VIII; Accedunt seorsim Theod. Janss. ab Almeloveen in C. Aurelianum notae et animadversiones*, ed. J. C. Amman (Amsterdam, 1709). [A]

CAELIUS AURELIANUS, *Medicinales responsiones* (*De salutaribus praeceptis*; *De significatione diaeticarum passionum*), ed. V. Rose (Anecdota Graeca et Graeco-latina, 2; Leipzig, 1870), 183-225.

CALCIDIUS, *Timaeus, a Calcidio translatus commentarioque instructus*, ed. J. H. Waszink (Plato Latinus, 4; London and Leiden, 1962).

CASSIUS THE IATROSOPHIST, *Problemata*, ed. J. L. Ideler (Physici et medici Graeci minores, i (Berlin, 1841; repr. Amsterdam, 1963), 144-67.

CELSUS, *De medicina*, ed. F. Marx (CML 1; 1915).

—— *La Préface du 'de medicina' de Celse*, ed. Ph. Mudry (Institut Suisse de Rome; Rome, 1982).

The Clementine Recognitions: *Die Pseudoklementinen Rekognitionen in Rufins Übersetzung*, ed. B. Rehm (Berlin, 1965).

GALEN, *De sectis ad introducendos*, ed. G. Helmreich, in *SM* iii. 1-32.

—— *De constitutione artis medicae ad Patrophilum*, ed. Kühn, 1. 224-304.

—— *De elementis ex Hippocratis sententia libri duo*, ed. G. Helmreich (Erlangen, 1878).

—— *De naturalibus facultatibus*, ed. G. Helmreich, in *SM* iii. 101-257.

—— *De usu partium*, ed. G. Helmreich, 2 vols. (Leipzig, 1907-9).

—— *De usu respirationis*, ed. D. J. Furley and J. S. Wilkie, in *Galen on Respiration and the Arteries* (Princeton, 1984).

—— *De atra bile*, ed. W. de Boer (CMG 5.4.1.1; Leipzig, 1937), 71-9.

—— *De placitis Hippocratis et Platonis*, ed. P. De Lacy (CMG 5.4.1.2, 2 parts; Berlin, 1978).

—— *De sanitate tuenda*, ed. K. Koch (CMG 5.4.2; Leipzig, 1923), 3-198.

—— *De morborum differentiis*, ed. Kühn, 6. 836-80.

—— *De plenitudine*, ed. Kühn, 7. 513-83.

—— *De tremore, palpitatione, convulsione et rigore*, ed. Kühn, 7. 584-642.

—— *De locis affectis*, ed. Kühn, 8. 1-452.

—— *De differentiis pulsuum*, ed. Kühn, 8. 493-765.

—— *De crisibus*, ed. B. Alexanderson, *Studia Graeca Latina Gothoburgensia* 23 (1967), 69-212.

—— *De diebus decretoriis*, ed. Kühn, 9. 769-941.

—— *De methodo medendi*, ed. Kühn, 10. 1-1021.

—— *De venae sectione adversus Erasistratum*, ed. Kühn, 11. 147-86.

—— *De simplicium medicamentorum temperamentis ac facultatibus*, ed. Kühn, 11. 379-892, 12. 1-377.

—— *De theriaca ad Pisonem*, ed. Kühn, 14. 210-94.

—— *In Hippocratis de natura hominis librum commentarii III*, ed. J. Mewaldt (CMG 5.9.1; Leipzig, 1914).

—— *In Hippocratis librum III Epidemiarum commentarii III*, ed. E. Wenkebach (CMG 5.10.2.1; Leipzig, 1936).

—— *In Hippocratis librum VI Epidemiarum commentarii VI*, ed. E. Wenkebach (CMG 5.10.2.2; Leipzig, 1940).

—— *Adversus ea quae a Iuliano in Hippocratis aphorismos enuntiata sunt libellus*, ed. E. Wenkebach (CMG 5.10.3; Leipzig, 1951), 33–70.

—— *De causis contentivis*, ed. M. C. Lyons (CMG Suppl. Or. 2; Berlin, 1969), 51–73, 131–41.

—— *De experientia medica*, ed. R. Walzer, *Galen on Medical Experience* (Oxford, 1944).

—— *De empirica subfiguratione*, ed. K. Deichgräber, in *Die griechische Empirikerschule* (Berlin, 1930; 2nd edn. Berlin and Zurich, 1965), 42–90.

PS.-GALEN, *Introductio sive medicus*, ed. Kühn, 14. 674–797.

—— *Definitiones medicae*, ed. Kühn, 19. 346–462.

—— *De historia philosophica*, ed. H. Diels, in *Dox.* 597–648.

HERO OF ALEXANDRIA, *Pneumatica*, ed. W. Schmidt, in *Heronis Alexandrini Opera*, i (Leipzig, 1899).

—— *Definitiones*, ed. J. L. Heiberg, in *Heronis Alexandrini Opera*, iv (Leipzig, 1912).

JOHN OF ALEXANDRIA, *Commentaria in librum de sectis Galeni*, ed. C. D. Pritchet (Leiden, 1982).

ORIBASIUS, *Collectiones medicae, Synopsis ad Eustathium filium*, ed. J. Raeder (CMG 6. 1–5; Leipzig, 1926–33).

PLINY, *Naturalis historia*, ed. C. Mayhoff (Leipzig, 1906–).

PLUTARCH, *Placita philosophorum*, in *Moralia*, ed. J. Mau, v/2.1 (Leipzig, 1971).

RUFUS OF EPHESUS, *Anatomy*, in *Œuvres de Rufus d'Éphèse*, ed. C. Daremberg and C. E. Ruelle (Paris, 1879; repr. Amsterdam, 1963), 168–85.

SEXTUS EMPIRICUS, *Opera*, ed. H. Mutschmann and J. Mau (Leipzig, 1914–58).

SIMPLICIUS, *Simplicii in Aristotelis Physicorum libros quattuor priores commentaria*, ed. H. Diels (Berlin, 1882).

SORANUS, *Gynaeciorum libri iv*, ed. J. Ilberg (CMG 4; Leipzig, 1927).

2. *Secondary Literature*

ALLBUTT, T. C. (1921), *Greek Medicine in Rome* (London).

ALPINUS, P. (1611), *De medicina methodica* (Padua).

BARNES, J., BURNYEAT, M., and SCHOFIELD, M. (edd.) (1982), *Science and Speculation: Studies in Hellenistic Theory and Practice* (Cambridge).

BÄUMKER, C. (1890), *Das Problem der Materie in der griechischen Philosophie* (Münster).

BENDZ, C. G. M. (1943), *Caeliana: Textkritische und sprachliche Studien zu Caelius Aurelianus* (Lunds Universitets Årsskrift, 38; Lund).

—— (1954), *Emendationen zu Caelius Aurelianus* (Skrifter utgivna av vetenskaps-societen i Lund, 44; Lund).

—— (1964), *Studien zu Caelius Aurelianus und Cassius Felix* (Skrifter utgivna av vetenskaps-societen i Lund, 55; Lund).

BERGHOFF, E. (1947), *Entwicklungsgeschichte des Krankheitsbegriffs* (Vienna).

BIGNONE, E. (1940), 'La dottrina epicurea del "clinamen"', *Atene e Roma*, 3rd ser. 8: 159–98.

BLUMENBACH, J. F. (1786), *Introductio in historiam medicinae litterariam* (Göttingen).

BOERHAAVE, H. (1751), *Methodus studii medici, emaculata et accessionibus locupletata ab Alberto ab Haller* (Amsterdam).

BURDACH, C. F. (1800), *Asklepiades und John Brown: Eine Parallele* (Leipzig).

CASTIGLIONI, A. (1947), *A History of Medicine*, trans. E. B. Krumbhaar (New York and London).

COCCHI, A. (1758), *Discorso primo sopra Asclepiade* (Florence).

DEICHGRÄBER, K. (1965), *Die griechische Empirikerschule: Sammlung der Fragmente und Darstellung der Lehre*, 2nd edn. (Berlin and Zurich).

DIELS, H. (1893), 'Über das physikalische System des Straton', *Sitzungsberichte d. Berliner Akademie*, 101–27.

—— (1969), *Kleine Schriften zur Geschichte der antiken Philosophie* (Darmstadt).

DIETZ, F. R. (1834), *Scholia in Hippocratem et Galenum*, i (Königsberg).

DOBSON, J. F. (1926–7), 'Erasistratus', *Proceedings of the Royal Society of Medicine* 20: 825–32.

DRABKIN, M. F. and I. E. (1951), *Caelius Aurelianus 'Gynaecia'* (Supplements to the *Bulletin of the History of Medicine*, 13; Baltimore).

DUMINIL, M.-P. (1983), *Le Sang, les vaisseaux, le cœur dans la collection hippocratique* (Paris).

EDELSTEIN, L. (1967), *Ancient Medicine*, ed. O. and C. L. Temkin (Baltimore).

EUCKEN, C. (1983), 'Zur Frage einer Molekulartheorie bei Herakleides und Asklepiades', *Museum Helveticum* 40: 119–22.

FISCHER, K.-D. (1982), 'Der Weg des Urins bei Asklepiades von Bithynien und in der Schrift "de opificio dei" des Kirchenvaters Lactantius', *Mémoires III du Centre Jean Palerne* (St-Étienne), 43–53.

FREDE, M. (1982), 'The Method of the So-called Methodical School of Medicine', in Barnes *et al.* (1982).

—— (1985), *Galen: Three Treatises on the Nature of Science* (Indianapolis).

—— (1987), *Essays in Ancient Philosophy* (Oxford).

FRITZSCHE, R. A. (1902), 'Der Magnet und die Athmung in antiken Theorien', *Rheinisches Museum* 57: 363–91.

FUCHS, R. (1892*a*), *Erasistratea quae in librorum memoria latent congesta enarrantur* (Diss. Berlin).

—— (1892*b*), 'Die Plethora bei Erasistratos', *Neue Jahrbücher für Philologie und Pädagogik* 38: 679–91.

—— (1894*a*), 'Anecdota medica Graeca', *Rheinisches Museum* 49: 532–58.

—— (1894*b*), 'De Erasistrato capita selecta', *Hermes* 29: 171–203.

—— (1903), 'Aus Themisons Werk über die acuten und chronischen Krankheiten', *Rheinisches Museum* 58: 67–114.

FURLEY, D. J. (1989), 'Strato's Theory of the Void', in *Cosmic Problems* (Cambridge).

—— and WILKIE, J. S. (1984), *Galen on Respiration and the Arteries: An Edition with English Translation of* De usu respirationis, An in arteriis natura sanguis contineatur, De usu pulsuum *and* De causis respirationis (Princeton).

GATZEMEIER, M. (1970), *Die Naturphilosophie des Strato von Lampsakos: Zur Geschichte des Problems der Bewegung im Bereich des frühen Peripatos* (Meisenhiem am Glan).

GIGANTE, M. (1975), 'Philosophia medicans in Filodemo', *Cronache ercolanesi* 5: 53–61.

GOTTSCHALK, H. B. (1961), 'The Authorship of *Meteorologica*, Book IV', *Classical Quarterly*, NS 11. 67–79.

—— (1965), 'Strato of Lampsacus: Some Texts', *Proceedings of the Leeds Philosophical and Literary Society* 11/6: 95–182.

—— (1980), *Heraclides of Pontus* (Oxford).

GUMPERT, C. G. (1794), *Asclepiadis Bithyniae Fragmenta* (Weimar).

GUTHRIE, W. K. C. (1965), *A History of Greek Philosophy* ii. *The Presocratic Tradition from Parmenides to Democritus* (Cambridge).

HALLER, A. VON (1776), *Bibliotheca medicinae practicae* i (Berne and Basle).

HARIG, G. (1983), 'Die philosophischen Grundlagen des medizinischen Systems des Asklepiades von Bithynien', *Philologus* 127: 43–60.

HARRIS, C. R. S. (1973), *The Heart and the Vascular System in Ancient Greek Medicine: From Alcmaeon to Galen* (Oxford).

HEIDEL, W. A. (1909), 'The ἄναρμοι ὄγκοι of Heraclides and Asclepiades', *Transactions of the American Philological Association* 40: 5–21.

HICKS, R. A. (1962), *Stoic and Epicurean* (New York).

HORNE, R. A. (1963), 'Atomism in Ancient Medical History', *Medical History* 7: 317–29.

JOUANNA, J. (1961), 'Présence d'Empédocle dans la collection hippocratique', *Bulletin de l'Association Guillaume Budé* 452–63.

LASSWITZ, K. (1879), 'Die Erneuerung der Atomistik in Deutschland durch Daniel Sennert und sein Zusammenhang mit Asklepiades von Bithynien', *Vierteljahresschrift für wissenschaftliche Philosophie* 3: 408–34.

—— (1890), *Geschichte der Atomistik vom Mittelalter bis Newton* (Hamburg and Leipzig).

LE CLERC, D. (1702), *Histoire de la Médecine* (Amsterdam).

LLOYD, G. E. R. (1966), *Polarity and Analogy* (Cambridge).

—— (1975), 'Alcmaeon and the Early History of Dissection', *Sudhoffs Archiv* 59: 113–47.

—— (1983), *Science, Folklore and Ideology* (Cambridge).

—— (1987), *The Revolutions of Wisdom* (Berkeley).

LONG, A. A., and SEDLEY, D. N. (1987), *The Hellenistic Philosophers* i. *Translations of the Principal Sources with Philosophical Commentary* (Cambridge).

LONGRIGG, J. (1976), 'The Roots of All Things', *Isis* 67: 420–38.

LONIE, I. M. (1964), 'The ἄναρμοι ὄγκοι of Heraclides of Pontus', *Phronesis* 9: 156–64.

—— (1965), 'Medical Theory in Heraclides of Pontus', *Mnemosyne*, 4th ser. 18: 126–43.

—— (1981*a*), *The Hippocratic Treatises 'On Generation', 'On the Nature of the Child', 'Diseases IV'* (Berlin).

—— (1981*b*), 'Hippocrates the Iatromechanist', *Medical History* 25: 113–50.

LÜCK, W. (1932), *Die Quellenfrage im 5. und 6. Buch des Lukrez* (Diss. Breslau).

MARELLI, C. (1981), 'La medicina empirica ed il suo sistema epistemologico', in *Lo scetticismo antico* (Elenchos: Collona di testi e studi sul pensiero antico, 6. 2), 657–76.

MEYER-STEINEG, T. (1916), *Das medizinische System der Methodiker* (Jena).

MILLER, H. (1957), 'The Flux of the Body in Plato's "Timaeus"', *Transactions of the American Philological Association* 88: 103-13.

—— (1962), 'The Aetiology of Disease in Plato's "Timaeus"', *Transactions of the American Philological Association* 93: 175-87.

MORAUX, P. (1977), 'Unbekannte Galen-Scholien', *Zeitschrift für Papyrologie und Epigraphik* 27: 1-63.

MOURELATOS, A. P. D. (1987), 'Quality, Structure and Emergence in Later Presocratic Philosophy', *Boston Area Colloquium in Ancient Philosophy* 127-94.

NEUBURGER, M. (1910), *History of Medicine*, trans. E. Playfair (London).

PIGEAUD, J. (1981*a*), *La Maladie de l'âme* (Paris).

—— (1981*b*), 'La physiologie de Lucrèce', *Revue des études latines* 58: 176-200.

—— (1982), 'Pro Caelio Aureliano', *Mémoires III du Centre Jean Palerne* (St-Étienne), 105-17.

PUSCHMANN, T. (1878), *Alexander von Tralles* (Vienna; repr. Amsterdam, 1963).

RAWSON, E. (1982), 'The Life and Death of Asclepiades of Bithynia', *Classical Quarterly*, NS 32: 358-70.

—— (1985), *Intellectual Life in the Late Roman Republic* (London).

RAYNAUD, A. G. M. (1862), *De Asclepiade Bithyno medico ac philosopho* (Diss. Paris).

REYMOND, A. (1927), *History of the Sciences in Greco-Roman Antiquity* (London).

RUBINSTEIN, G. L. (1985), 'The Riddle of the Methodist Method: Understanding a Roman Medical Sect' (Diss. Cambridge).

SCARBOROUGH, J. (1969), *Roman Medicine* (London).

SCHMEKEL, A. (1938), *Die positive Philosophie in ihrer geschichtlichen Entwicklung* (Forschungen zur Philosophie des Hellenismus, 1; Berlin).

SCHÖNER, E. (1964), *Das Viererschema in der antiken Humoralpathologie* (*Sudhoffs Archiv*, Beih. 4).

SEDLEY, D. N. (1982), 'Two Conceptions of Vacuum', *Phronesis* 27: 175-93.

SINGER, C., and UNDERWOOD, E. A. (1962), *A Short History of Medicine* (Oxford).

SMITH, W. D. (1979), *The Hippocratic Tradition* (Ithaca and London).

SMITH, W. D. (1982), 'Erasistratus's Dietetic Medicine', *Bulletin of the History of Medicine* 56: 398–409.

SOLMSEN, F. (1956), 'On Plato's Account of Respiration', *Studi italiani di filologia classica* 27–8: 544–8.

—— (1961), 'Greek Philosophy and the Discovery of the Nerves', *Museum Helveticum* 18: 150–67, 169–97.

SPRENGEL, K. (1792), *Versuch einer pragmatischen Geschichte der Arzneykunde* (Halle).

STADEN, H. VON (1975), 'Experiment and Experience in Hellenistic Medicine', *Bulletin of the Institute of Classical Studies* 22: 178–99.

—— (1976), 'Hairesis and Heresy: The Case of the *haireseis iatrikai*', in B. F Meyer and E. P. Sanders (edd.), *Jewish and Christian Self-Definition* iii (London), 76–206.

—— (1989), *The Art of Medicine in Early Alexandria: Herophilus and his School* (Cambridge).

STÜCKELBERGER, A. (1972), 'Lucretius reviviscens: Von der antiken zur neuzeitlichen Atomphysik', *Archiv für Kulturgeschichte* 54: 1–25.

—— (1974), 'Empiriker Ansätze in der antiken Atomphysik', *Archiv für Kulturgeschichte* 56: 124–40.

—— (1979), *Antike Atomphysik: Text zur antiken Atomlehre und ihren Wiederaufnahme in der Neuzeit* (Munich).

—— (1984), *Vestigea Democritea: Die Rezeption der Lehre von den Atomen in der antiken Naturwissenschaft und Medizin* (Schweizerische Beiträge zur Altertumswissenschaft, 17, 17).

SUSEMIHL, F. (1892), *Geschichte der griechischen Litteratur in der Alexandrinerzeit* ii (Leipzig).

TAYLOR, A. E. (1928), *A Commentary on Plato's* Timaeus (Oxford).

THESLEFF, H. (1961), *An Introduction to the Pythagorean Writings of the Hellenistic Period* (Åbo).

TOURTELLE, E. (1804), *Histoire philosophique de la médecine, depuis son origine jusqu'au commencement du 18ème siècle* i (Paris).

VALLANCE, J. T. (1988), 'Theophrastus and the Study of the Intractable: Scientific Method in "De lapidibus" and "De igne"', in *Theophrastean Studies*, ed. W. W. Fortenbaugh and R. W. Sharples (Rutgers Studies in Classical Humanities, 3; New Brunswick), 25–40.

VIANO, C. A. (1981), 'Lo scetticismo antico e la medicina', in *Lo scetticismo antico* (Elenchos: Collana di testi e studi sul pensiero antico, 6. 2), 563–656.

VOSS, O. (1896), *De Heraclidis Pontici vita et scriptis* (Rostock).

WELLMANN, M. (1895), *Die pneumatische Schule bis auf Archigenes in ihrer Entwicklung dargestellt* (Philologische Untersuchungen 14; Berlin).

— (1900), 'Zur Geschichte der Medizin im Alterthum', *Hermes* 35: 349–94.

— (1901), *Die Fragmente der sikelischen Ärzte Akron, Philistion und des Diokles von Karystos* (Berlin).

— (1905), 'Herodots Werk περὶ τῶν ὀξεῶν καὶ χρονίων νοσηματων', *Hermes* 40: 580–604.

— (1908), 'Asklepiades aus Bithynien von einem herrschenden Vorurteil befreit', *Neue Jahrbücher* 21: 684–703.

— (1913), *A. Cornelius Celsus: Eine Quellenuntersuchung* (Philologische Untersuchungen 23; Berlin).

— (1922), 'Der Verfasser des Anonymus Londinensis', *Hermes* 57: 396–429.

— (1929), 'Spuren Demokrits von Abdera im Corpus Hippocraticum', *Archeion* 11: 297–330.

WRIGHT, M. R. (1981), *Empedocles: The Extant Fragments* (New Haven).

ZELLER, E. (1909), *Die Philosophie der Griechen in ihrer geschichtlichen Entwicklung*, 5th edn., iii (Leipzig).

Index Locorum

This index includes passages discussed in my text, and not those simply cited as supporting evidence.

General Index